SpringerWienNewYork

Thomas Jakl, Petra Schwager (Eds.)

Chemical Leasing Goes Global

Selling Services Instead of Barrels:
A Win-Win Business Model for Environment and Industry

SpringerWienNewYork

Thomas Jakl

Federal Ministry of Agriculture, Forestry,
Environment and Water Management, Vienna, Austria

Petra Schwager

United Nations Industrial Development Organization,
Vienna International Centre, Vienna, Austria

Printed in Austria

SpringerWienNewYork is part of
Springer Science+Business Media
springer.at

Cover design: United Nations Office at Vienna, PRES
Typesetting: Grafik Rödl, 2486 Pottendorf, Austria
Printing: Holzhausen Druck und Medien GmbH, 1140 Wien, Austria

Printed on acid-free and chlorine-free bleached paper
SPIN: 12040322

With 82 partly coloured Figures

Library of Congress Control Number: 2007937034

ISBN 978-3-211-73751-4 SpringerWienNewYork

Contents

Acknowledgements

The Federal Ministry of Agriculture, Forestry, Environment and Water Management and United Nations Industrial Development Organization as initiator of this book, would like to thank the following companies and persons, whose personal commitment contributed substantially to the success of this book project:

Eva-Maria Reiss for managing this book project.

Russia
State Unitary Enterprise "Vodokanal of Saint Petersburg"
Service Company ERG Ltd.
Henkel-ERA AG
Henkel Moscow AG
Federal State Unitary Enterprise "Krasny Bor"
Limited Company "Fuel Systems"
R&D Company RIVT
Service Company "ECOVIP Systems"
Federal State Unitary Enterprise "Admiralteiskie Verfi"
Public Corporation "Electrotechnical Plant"
Public Corporation "FERRO-PRIBOR"
State Unitary Enterprise "GOSMETR"

Professor, Doctor Vladimir Alexandrovich GRACHEV, Chairman of the State Duma Environmental Committee, Deputy of Russian Parliament, Corresponding Member of Russian Academy of Science, Moscow.
Professor Kirill Yakovlevich KONDRATYEV, Academician, Counsellor of the Russian Academy of Sciences, First Chairman of the Board of the North-Western International Cleaner Production and Environmental Management Centre, St. Petersburg.
Ekaterina Dmitrievna BELYAKOVA, Head of Department for External Economic Relations and Regional Cooperation, Government of Saint-Petersburg Committee for External Relations.
Doctor Dmitry Alexeevich GOLUBEV, Chairman of the Committee of Nature Use, Environmental Protection and Ecological Safety, Government of St. Petersburg.

Egypt
Egyptian Ministry of Trade and Industry (MTI)
The Egypt National Cleaner Production Center
Akzo Nobel Egypt'

ABB ARAB,
Dr. Badawi Chemical Work
General Motors Egypt

Mexico
Chemical Mac Oil S.A. de C.V.
Fideicomiso Ingenio de san Cristobal 80333 S.A. de c.V.
Mardi Inc. S.A.de C.V.
Cromadora Delgado, S.A. de C.V.

Daniela Jäger-Bergaus, MsC, Environmental Manager,
Borealis Polyolefine GmbH

Foreword

In the last years the Austrian Ministry of Environment made every effort to promote innovative ways of chemicals management based on resource efficiency as well as on precaution. Applying „Chemical Leasing“, profits do not result from the quantity of chemicals sold, but rather from the services rendered by means of the chemicals. If business interests are no longer related to the sale of chemical substances it is in the interest of all the parties involved, to use the substances with maximum efficiency. This clearly is the environmental policy motivation for this new focus, which ensures that economic success and ecological benefits are no longer contradictory.

In 2003, we published our first book concerning the new business model and its advantages: "Chemical Leasing – An Intelligent and Integrated Business Model with a View to Sustainable Development in Materials Management". While this first comprehensive publication dealt with the theory behind Chemical Leasing and highlighted both the economic and ecological potentials of service-oriented models in the chemicals industry, this book before you shows also the first successful results of the cooperation between the Austrian Ministry of Environment and UNIDO. UNIDO picked up the idea of "Chemical Leasing" and implemented it on an international scale. Together with UNIDO's National Cleaner Production Programme we are now able to present these very promising international results along with those derived from Austrian pilot projects.

This book also covers the important connection of the new EU's chemicals policy, REACH, and "Chemical Leasing". Within REACH, cooperation, networking and communication based on documentation, evaluation and minimisation of hazards is indispensable. The new business model can serve as the basis for this intensified dialogue and cooperation.

Chemical Leasing found its way into the political discussion on European and international level, into contributions to academic work and into applied research programmes and most importantly Chemical Leasing triggered a more than promising trend within the chemical-related business sectors. This book impressively documents that this concept launched a real success story both in terms of environmental and economic indicators thus giving the political term "sustainable chemical management" not only a concrete meaning but also a prosperous perspective.

Josef Pröll
Federal Minister of Agriculture, Forestry,
Environment and Water Management

Preface

Open trade, international investment flows, information technology and global environmental concerns are major elements of the current multilateral order. New integrated concepts have to be applied to ensure progressive and environmentally sound development. These concepts will be based on private sector-led productivity gains in developing countries and countries with economies in transition.

Since 1994, the United Nations Industrial Development Organization (UNIDO) has been operating the Cleaner Production programme. The goal of this programme is to assist small and medium-sized enterprises (SMEs) in developing countries and countries with economies in transition to improve their productivity while preventing unfavourable impacts on the environment.

Encouraging innovative concepts that respond to new challenges constitutes an integral element of UNIDO's Cleaner Production strategy. Chemical Leasing is one such important and innovative concept. It is a novel service-based business model that supports sustainable chemicals management and responds to the latest changes in chemical policies. It also aims to achieve the objectives of the voluntary international accord known as Strategic Approach to International Chemicals Management (SAICM). Chemical Leasing provides practical solutions for industry to become more efficient. At the same time it reduces the unnecessary consumption of hazardous chemicals and protects human health and the environment. This represents an integrated preventive environmental strategy and a clear win-win situation for industry and the environment.

In 2004, UNIDO and the Austrian Ministry of Environment decided to join hands to support Chemical Leasing with the launch of a global project. This project is currently being implemented in close cooperation with UNIDO's National Cleaner Production Centres in Egypt, Mexico and Russia. This book presents the encouraging results of this challenging undertaking, as well as the new instruments developed by UNIDO to facilitate the efficient implementation of Chemical Leasing.

With this joint publication UNIDO and the Austrian Government wish to share with the global community the benefits that may be derived from Cleaner Production and Chemical Leasing business models for the economy, the environment and risk management. In doing so, we hope to encourage an increasing number of enterprises to adopt similar initiatives to foster sound chemicals management at the global level.

Kandeh Yumkella
Director-General of the United Nations Industrial
Development Organization

Chapter 1

Chemical Leasing – An Introduction

Thomas Jakl

Chemical products are of vital importance to the economic and social well being of our society in terms of trade and employment. Numerous chemical substances are used in a broad range of products and processes. They also help to improve the health and quality of life of the population through new developments such as in pharmaceuticals, food safety, pesticides and in many other products. During or after their use, these substances are released to the environment to a varying extent and may lead to pollution. This results in both ecological and – due to unnecessary losses – economic disadvantages.

While it is vital that the chemical industry itself remains healthy, it is equally important that it contributes in a sustainable way to social and economic developments. This means that the impact of chemicals on the environment and human health has to be minimised. Only this approach will ensure that the chemical industry remains competitive and gains the confidence of the general public and policy-makers.

Taking a closer look at the multilateral instruments developed over the last decades in chemicals policy, the following interpretation applies:

Some instruments focussed on a specific group of substances, grouped according to their inherent properties, as it was the case within the Montreal Protocol to the Vienna Convention (substances regulated have the property of destroying the ozone layer) or as it was the case within the Stockholm Convention (substances regulated have POP-characteristics). Another possible approach of grouping was to choose the specification of uses and the application of risk management measures as criteria for inclusion in a multilateral agreement as it was the case with the Rotterdam Convention, dealing with industrial chemicals and pesticides.

In a nutshell: Specific environmental problems triggered substance – specific measures.

The subsequent application of risk assessment and risk management measures was consensually put into legal language in order to achieve the goal of *"using the right chemicals for various purposes"*. This was an adequate – and to a great extent successful – approach to tackle problems caused by specific cause-and-effects phenomena thus representing the traditional paradigm of environmental policy. Within this classical traditional paradigm the relationship between chemical policy and the chemical industry is coined by diverging rationales and targets. The presence of synthetic chemical substances in air, water and soil led to contamination of ecosystems and at the same time establishes a risk for human health. The political aim for reducing exposure of humans and the environment was linked to less profit for the economic sector as it was equally linked to reduced use and turnover of chemical products.

Sustainable material management has to pursue both economic and ecological objectives in order to succeed in global competition. Material flows have to be optimised in quantitative and qualitative terms so that the services in question can be rendered with maximum efficiency and minimal emissions. Moreover, business companies are generally aware of the fact that only those companies will survive that continuously improve their performance while enhancing customer orientation and, above all, avoiding excessive resource consumption.

Given that background, bridging the gap between the approaches of environmental policy and the parameters for success of the chemicals industry was the central challenge. Making efficient use of chemicals also a goal for companies whose economic success was directly coupled to increased consumption was the achievement of the impossible. A new business model should allow producers of chemicals to accept a reduced use of their product from an economic point of view. Selling a maximum amount of barrels is too narrow a concept to judge economic success if economic prosperity shall be embedded in a sustainable development. Therefore additional stages of the life cycle of chemical products have to be integrated into the factors of economic interests for chemical business. That means extending their commercial pattern from the "produce and sell" world to include the stage in which their product is applied. The only way to achieve that is to extend their economic interest to include the usage-phase of the chemical products: Business companies that originally produced cleaning agents have to be involved into the cleaning process and into the cleaning business if resource efficiency concerning their own product is to be part of their economic success. A paint producing company only will shift its economic interest toward "less product consumption" if the application and use of its product is part of its business thus turning "reduced use of paint" from bad news to good news even for paint producers.

As a consequence, potentials for solving this conflict of interest between chemicals industry and environment policy are mainly seen in promoting new business models in which profits do not result from the quantity of chemicals sold, but rather from the services rendered by means of the chemicals. The term "Chemical Leasing" stands for such a service-oriented business model in which companies producing chemicals are connected to companies applying chemicals. The chemicals needed for a specific service are not simply sold to the customer, but – in contrast to the traditional business model – are made available for use and are maintained. Within Chemical Leasing applications business interests are no longer related to the sale of chemical substances, but to the sale of chemical services, it is in the interest of all of the parties involved to use the substances with maximum efficiency. When partners opt for that business model, their economic interest also comprises the usage phase of the product making resource efficiency a high priority target for them. This built-in economic driver for most efficient use of chemicals clearly is the environmental policy motivation to promote Chemical Leasing.

Chemical Leasing and substitution

The replacement of chemicals by alternative substances or technologies is – as one option among many others – an inherent element of optimization and innovation. That continues to be so within the Chemical Leasing concept. There is no clear cut border line or contrast between the concept of Chemical Leasing and the option of substitution. Of course substitution is not a goal on its own and each alternative to an existing chemical has to be carefully judged in view of its complete life-cycle. According to our understanding the new business culture encompassing Chemical Leasing models leads to a transparent and open way of optimising products and processes. Replacing a chemical with another chemical or even with another technology of course is not a taboo in this repect and might be an option chosen by the partners involved.

Moreover continuous optimisation in economic and also ecologic terms is an inherent and integrated element of the Chemical Leasing concept according to our understanding. Giving a high priority to the substitution of chemicals is therefore no contradiction at all to the application of Chemical Leasing but might be a logical and natural result of the optimisation process.

"Chemical Leasing" is put into practice in Austria

The Austrian Federal Ministry of Agriculture, Forestry, Environment and Water Management (BMLFUW) commissioned various studies, showing that almost 4000 Austrian companies would basically qualify for the application of Chemical Leasing models, cutting today's annual use of 150,000 tons of chemicals by one third. On average, the user of such new business models can expect cost savings of approximately 15%.

The Ministry is supporting the establishment of pilot implementation projects in Austria, where six implementation projects covering a wide range of applications and technologies have been or are being implemented.

The Chemical Leasing model has won two awards so far: In December 2004 the "Genius 2004" award in the innovation sector and in early 2005 the "Constantinus" award in the sector of management consulting.

The first publication

In 2003, we published our first book concerning the new business model and its advantages: *"Chemical Leasing – An Intelligent and Integrated Business Model with a View to Sustainable Development in Materials Management"*. This first publication dealt with the theory behind Chemical Leasing and highlighted both the economic and ecological potentials of service-oriented models in the chemicals industry. Possible economic and environmental benefits as well as possible obstacles and drawbacks were elucidated. A common understanding about the nature and characteristics of those approaches, suitable to be seen as examples for function or service-oriented chemical business companies was developed.

As a further activity the Austrian Federal Ministry of Agriculture, Forestry, Environment and Water Management (BMLFUW) in co-operation with the OECD organised an international conference in Vienna in November 2003, which was called *"Experiences and Perspectives of Service-oriented Strategies in the Chemicals Industry and Related Areas"*. The conference was attended by around 100 participants from 16 countries. The conference gave an excellent overview about international approaches how to introduce service elements into the chemical business. The participants comprised representatives from authorities, the chemical industry, in particular down stream users, process engineers and economists along with representatives from academia.

In their presentations the experts confirmed the prospect of success of such new business models among which the Chemical Leasing concept seems to be the most advanced and comprehensive one showing the highest potentials both in terms of resource efficiency as well as environmental compatibility.

One of the most important results of the November 2003 Austrian conference was that the legal framework, within which the partners of such new business models act, is of high interest. Especially the legal nature of the chemical within the cycle of the leasing model is decisive. This should become a clear and intelligible factor.

2007 – a crucial step forward due to our co-operation with UNIDO

The experience of the Austrian pilot projects is now used for the UNIDO project "Promotion and implementation of Closing-the-Loops cooperation and business models in the chemical industry". Consultants who supported the Austrian pilot projects are now acting as international experts in Egypt, Mexico and Russia. This extremely successful UNIDO cooperation in particular led to a remarkable increase in applications of this model, developing it from a theoretical concept into a well established and promising perspective for both: producers and users of chemicals. UNIDO picked up the idea of "Chemical Leasing" and put it on an international basis. With the help of UNIDO's National Cleaner Production Programme we are now able to present the very promising international results along with our Austrian pilot projects in this book.

Cleaner Production and Chemical Leasing both are efficient policy instruments for sustainable chemicals management. The combination of Cleaner Production and Chemical Leasing has proven to be a win-win situation for the economy and the environment and can be applied to business companies of different sizes.

Chemical Leasing and REACH

This book also covers the connection between the new EU Chemicals Policy (REACH) and Chemical Leasing. Within REACH, cooperation, networking and communication based on documentation, evaluation and minimisation of hazards is indispensable. The new business model can serve as a basis for this intensified dialogue and cooperation.

Chemical Leasing – a "flagship initiative"

On the one hand it is a central aim of environment policy in the field of chemicals, to provide the adequate legal framework to ensure the necessary level of protection while at the same time providing instruments to strengthen the competitiveness of eco-efficient and environmentally friendly products and services. In accordance with this goal a couple of initiatives were launched by Austria involving all relevant interest groups. A mix of financial, institutional and technical instruments within the scope of introducing new technologies is being offered with a focus on communication and transfer of know-how. This aims at faster market penetration of products and technologies and business models with an enhanced ecological profile. The Chemical Leasing-initiative represents the flagship of instruments in Austria's chemical policy framework. We also try to integrate this concept at EU-level as well as in international fora.

When Austria held the Presidency of the European Union during the first half of 2006, chemicals policy was on top of the environmental agenda. Completing the council's work on the "common position" on REACH, leading the EU's negotiations on SAICM (Strategic Approach to International Chemicals Management) as well as on the Stockholm Convention on Persistent Organic Pollutants together with hosting a range of international workshops and conferences were cornerstones of our work. When SAICM was finalised during the International Conference on Chemicals Management in early 2006 (ICCM, Dubai, February 2006) the Austrian Federal Minister for the Environment, and the UNIDO Director General hosted an event committing themselves to Chemical Leasing as their common political priority.

Austria will continue to promote the Chemical Leasing concept and try to integrate its experience also into the future SAICM implementation processes as it ideally complies with the new paradigm of environmental policy and of chemicals policy in particular. A paradigm that incorporates measures addressing both – the properties of chemicals applied, as well as the quality of practices used in their application. We understand that the term "strategic" implies the establishment of a framework of measures ensuring optimised "eco-efficiency" thus minimising the release of chemicals into the environment. *"Use the right chemicals right"* should be the overall goal of this new approach.

One element within this possible "framework for eco-efficiency" should therefore be a catalogue of measures stimulating the dissemination of Best Practices as outlined in SAICM's "Global Plan of Action". Chemical Leasing shall become a central element of that instrument and its implementing processes.

Due to the Austrian initiatives and in particular due to the ambitious activities of UNIDO, service-oriented approaches represent a growing and promising tendency both within chemicals policy as well as within the chemicals industry. These models are eco-efficient by nature and show a tremendous potential in terms of achieving ecological as well as economic benefits.

It is far from exaggerating when we state that due to these experiences gained so far, Chemical Leasing is becoming a worldwide perspective of making sustainable development happen. The catalogue of examples and evaluations contained in this book shall pave the way to broad implementations of Chemicals Leasing while minimizing the economic risks for companies.

References

Ayres R (1989) Industrial metabolism. In: Ausubel J (ed) Technology and environment. National Academy Press, Washington, D.C.

Behrens A (2004) Environmental policy instruments for dematerialisation of the European Union. SERI Background Papers, No 7. Sustainable Europe Research Institute, Vienna

EUROSTAT (2001) Economy-wide material flow accounts and derived indicators. A methodological guide. Statistical Office of the European Union, Luxembourg

Giljum S (2006) Material flow-based indicators for evaluation of eco-efficiency and dematerialisation policies. In: Lawn P (ed) Sustainability indicators in Ecological Economics. Edward Elgar, Cheltenham

Giljum S, Eisenmenger N (2004) North-South trade and the distribution of environmental goods and burdens: a biophysical perspective. Journal of Environment and Development 13(1): 73–100

Giljum S, Hak T, Hinterberger F, Kovanda J (2005) Environmental governance in the European Union: strategies and instruments for absolute decoupling. International Journal for Sustainable Development 8(1/2): 31–46

Hafner G, Jakl T, Loibl G (2002) Conference conclusions by the chairs. In: Freytag E, Jakl T, Loibl G, Wittmann M (ed) The role of precaution in chemicals policy. Favorita Papers, Diplomatic Academy of Vienna

Hinterberger F, Hammer M, Luks F, Omann I, Stewen M, Stocker A, Hutterer H, Strigl A, Schmidt-Bleek F (2004) Eco-Efficient Innovation. State of the Art and Policy Recommendations. Paper for the Informal Environment Council on 16–18 July 2004 in Maastricht

Jacobsson N (2006) Emerging product strategies – selling services of remanufactured products (IIIEE Dissertation 2000: 6). Lund University, Sweden

Jakl T, Joas R, Nolte R, Schott R, Windsperger A (2004) Chemical Leasing – an intelligent and integrated business model with a view to sustainable development in materials management. Springer, Wien New York

Joas R, Schott R (2004) R: Promoting and inhibiting factors, chap. 8, op. cit.

Mont O (2001) Introducing and developing a Product-Service System (PSS) concept in Sweden (IIIEE Reports 2001: 6). The International Institute for Industrial Environmental Economics, Lund University, Sweden

Nolte R (2004a) The six key questions on Chemical Leasing, chap. 2, op. cit.

Nolte R (2004b) Chemical Leasing in Austria, chap. 6, op. cit.

OECD (2004) Recommendation on material flows and resource productivity. Adopted by the OECD Council on 21 April 2004. No. OECD, Paris

Oldham J, James P, Shaw B (2003) Delivering resource productivity: the service solution. Green Alliance, London

Perthen-Palmisano B, Jakl T (2004) Chemical Leasing – cooperative business models for sustainable chemicals management; Science and Pollution Research (Vol 12, No 1/ Januar 2005 ecomed publishers), Landsberg, Tokyo Mumbai Seoul Melbourne Paris

Pranshu S (2003) Transition to chemical management services, opportunities, barriers and drawbacks in a European Context (IIIEE Master's Theses 2003: XX). The International Institute for Industrial Environmental Economics, Lund University, Sweden

Spangenberg J, Lorek S (2002) Environmentally sustainable household consumption: from aggregate environmental pressures to priority fields of action. Ecological Economics 43(2–3): 127–140

UBA (2004) UBA-Germany, Report on the International Workshop on Sustainable Chemistry, Dessau 27 to 29 January 2004, Session V. Implementation <http://www.sustainable-chemistry.com/download/GESAMTREPORT.pdf>

UNIDO (2004) UNIDO, UNIDO Information Document on Cleaner Production (BAT/BEP) for SAICM PrepCom 2, Vienna, 9 June 2004SAICM/PrepCom.2/INF/4

Windsperger A (2004) The variants of Chemical Leasing, chap. 4, op. cit.

Chapter 2

Chemical Leasing and Cleaner Production

Petra Schwager

Background

Globalisation, changing patterns of consumption and production and increased environmental concerns are creating new challenges for companies in developing countries and countries with economies in transition to stay competitive in the global market. Integrated and multidisciplinary approaches are required to face the dramatic rises in the intensive use of natural resources and over-reliance on the ability of the natural environment to absorb the wastes generated by economic activity.

To address the challenges of the new global context and to enhance economy wide productivity in a sustainable manner, the United Nations Industrial Development Organisation (UNIDO) focuses its activities on three thematic priorities: poverty reduction through productive activities, trade capacity building and environment and energy.

The organisation's worldwide Cleaner Production Programme is an important tool to bridge these three priorities and plays a fundamental role in promoting sustainable industrial development and sound chemicals management in developing countries and countries with economies in transition.

The Cleaner Production Programme

The Cleaner Production concept was adopted by the United Nations Conference on Environment and Development (UNCED) in Rio de Janeiro, Brazil in 1992 as an efficient means to achieve sustainable industrial development. It is an integrated preventive environmental strategy to increase resource efficiency and to reduce risks to humans and the environment. It aims at minimising the generation of pollution and waste at source, rather than addressing and mitigating just the symptoms by only technically "dealing with" an existing waste/pollution problem. Cleaner Production can be applied to production processes, products and services.

In the early nineties of the last century UNIDO together with UNEP (United Nations Environment Programme) launched a joint National Cleaner

Production Centres Programme to contribute to sustainable productivity and to the competitiveness of industries in developing countries and countries with economies in transition.

Over the past decade UNIDO, being responsible for the implementation and management of the Programme, established National Cleaner Production Centres (NCPCs) and National Cleaner Production Programmes (NCPPs) in 38 countries. In addition, UNIDO started its first Regional Cleaner Production Network in Latin America in 2005 with 14 countries participating.

The Cleaner Production Programme provides an active contribution to the implementation of the United Nations Millennium Declaration and the achievement of United Nations Millennium Development Goals 1 (eradicate extreme poverty and hunger), 7 (ensure environmental sustainability) and 8 (develop a global partnership for development).

It is based on a multi-disciplinary and multi-stakeholder approach, aiming at involving the different levels of industry, government, academia and the financial sector. NCPCs and NCPPs build on existing structures. They are established in national institutions that are close to industry and can support the build-up of national capacity in Cleaner Production, e.g. universities, industrial chambers and associations, research centres.

Through their cooperation with the main national key stakeholders, the NCPCs and NCPPs are setting up CP networks that promote the concept in the respective countries and ensure its adjustment to the local conditions. To foster the sustainable application of Cleaner Production, NCPCs and NCPPs carry out five core sets of activities: in-plant assessments to demonstrate the benefits of CP at company level; national capacity building to train national CP experts, company staff, government and academia; development and transfer of Environmentally Sound Technologies (ESTs); promotion of CP and EST business cooperation and investment projects; dissemination of information on CP, EST and related emerging topics; and CP policy advice to assist governments in creating mechanisms and instruments to support CP. All these activities are interrelated and aim at increasing the productivity and competitiveness of the selected priority industrial sectors in the host countries.

Cleaner Production and Sustainable Industrial Resource Management

A critical element of Cleaner Production is that it results in a "win-win" scenario for industry and the environment as it implies striving for continuous resource efficiency to create economic savings for the company. In this way it enhances the competitiveness of industry, promoting sustained social advancement in a way which is compatible with environmental protection.

Based on its experience in applying CP to industry at a global level, UNIDO further developed UNEP's CP concept and in 2002 launched the **Holistic CP approach** that emphasises the company and sectoral level and takes into account the entire product cycle.

At company level, the activities should progressively go beyond process improvement to take the entire product cycle into account, from raw material production to product recycling and/or disposal and involving all functions and departments of the company as well as suppliers and customers. This holistic approach allows the companies to design, produce and promote innovative, environmentally and economically sound products, improving their ability to successfully negotiate their position in the global markets.

At the sectoral level, national sector-specific CP strategies should be developed. This would allow CP interventions at company level to be put into a broader perspective and would point out possible synergies between companies, suppliers and customers, allowing gains obtained at the sector and national levels to be optimised. It would also enable adequate choices in technology development and transfer in the country as well as the preparation of coherent investment plans to be presented to potentially interested companies and financial institutions (Fig. 1).

UNIDO recently developed its new **Sustainable Industrial Resource Management** (SIRM) strategy. Unlike the traditional approaches to environmental management, the SIRM concept promotes the idea of achieving sustainable industrial development through the implementation of circular material and energy flows in the entire production chain and the reduction of the amount of material and energy used by means of solutions for greater efficiency. It focuses on the idea that total materials cycles can be optimised and modelled on the self-sustaining cycles of nature. The SIRM concept abandons the "cradle to grave" principle of linear product cycles and aims at effectively closing the material and energy loops. The above-indicated objectives are expected to be achieved through the following means:

- Separation of the material flows into two parts: technical flow (including non-biodegradable materials) and nutrients flow (with organic and biodegradable materials) to facilitate internal recycling and reuse;
- Substitution of non-renewable energy sources by renewable ones.

In addition, the SIRM approach seeks to develop new models to encourage a shift from selling products to supplying services, by this way modifying the supplier/user relationship and resulting in a win-win situation for the economy

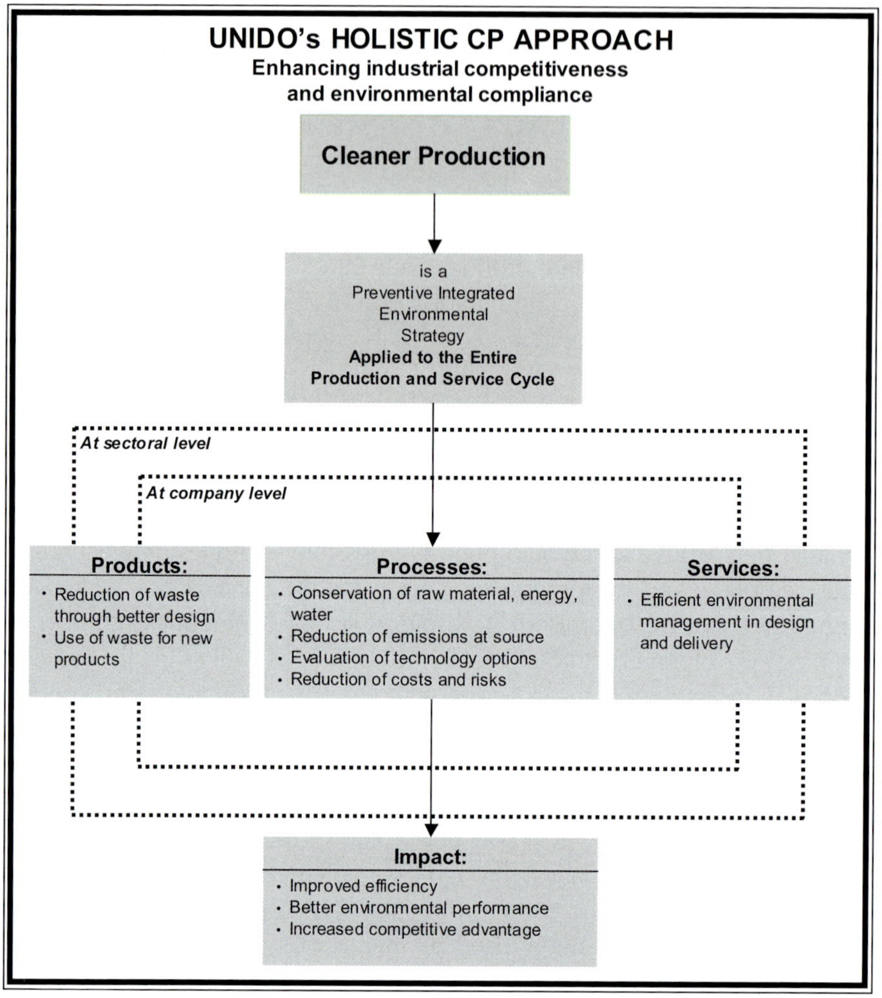

Figure 1

and the environment. Chemical Leasing is an example of this type of a new service-oriented business model.

Chemical Leasing – a service based business model to support sound chemicals management

The concept of Chemical Leasing is based on the preventive idea of Cleaner Production and represents an important aspect of UNIDO's SIRM approach. It is a shift from the traditional business concept that focuses on a constant increase in sales volume towards a more service and value-added approach. Chemical Leasing business models provide concrete solutions to the effective management of chemicals and to negative releases to the environment.

In 2004, UNIDO's Cleaner Production Programme and the Austrian Federal Ministry of Agriculture, Forestry, Environment and Water Management (BMLFUW), supported by the Austrian Federal Ministry for European and International Affairs (BMEIA) launched the joint project "Promotion and implementation of closing-the-loops cooperation and business models in the chemical industry". The main aim of this global undertaking is to demonstrate the applicability of Chemical Leasing business models in developing countries and countries with economies in transition and to further develop and promote the concept of Chemical Leasing.

Core elements of the joint project are national capacity building and Chemical Leasing demonstration projects in selected industries. These activities are being implemented in Egypt, Mexico and the Russian Federation in close cooperation with the respective National Cleaner Production Centres. The selected NCPCs have already established good cooperation with the chemical industry and are well familiar with the problems associated with sustainable industrial development in their countries. Besides offering logistical and administrative support, the Centres deliver important information on the national chemical sector and identify local companies and other relevant stakeholders. The role of the Centres in coordinating the implementation and monitoring of Chemical Leasing business models at the national level is extremely important.

The results obtained in the three countries are very encouraging and are outlined in Chapter 5 of this publication.[1]

At a global level, UNIDO seeks to further promote and develop the concept of Chemical Leasing. For this purpose, an **international multi-stakeholder working group on Chemical Leasing** that includes representatives from government, industry, certification bodies, academia and the consultant sector was launched. In close cooperation with this working group – which meets on a regular basis – UNIDO developed the following definition of Chemical Leasing.

1 The outlined projects reflect the results obtained until spring 2007.

UNIDO Definition

Chemical Leasing

- *Chemical Leasing is a service-oriented business model that shifts the focus from increasing sales volume of chemicals towards a value-added approach. The producer mainly sells the functions performed by the chemical and functional units[2] are the main basis for payment.*
- *Within Chemical Leasing business models, the responsibility of the producer and service provider is extended and may include the management of the entire life cycle.*
- *Chemical Leasing is a win-win situation. It aims at increasing the efficient use of chemicals while reducing the risks of chemicals and protecting human health. It improves the economic and environmental performance of participating companies and enhances their access to new markets. Key elements of successful Chemical Leasing business models are proper benefit sharing, high quality standards and mutual trust between participating companies.*

Chemical Leasing builds on the concept of the Cleaner Production approach as it is a preventive, environmental, service-oriented business model that aims at improving overall efficiency, enhancing environmental performance and increasing the competitive advantage of industry.
Based on experiences obtained in Egypt, Mexico and Russia and to enable a systematic approach of the implementation of Chemical Leasing business models at company level, UNIDO has developed a first toolkit. It covers the main steps to be undertaken to ensure the smooth and efficient application of Chemical Leasing in the industries of different countries. An overview of the general content of UNIDO's **Chemical Leasing toolkit** is included in Chapter 8.

An important lesson learned from the implementation of the project "Promotion and implementation of closing-the-loops cooperation and business models in the chemical industry" is the need for quality assurance of Chemical Leasing business models. The models must result in improved environmental performance of the participating companies, reduced risks of chemicals and enhanced human health protection.

2 Functions performed by a chemical might include: number of pieces cleaned; amount of area coated, etc.

Chemical Leasing and Cleaner Production – efficient policy instruments for sustainable chemicals management

Experience has shown that Chemical Leasing and Cleaner Production support sustainable chemicals management, respond to the changes in chemical policies and help to achieve the objectives of the voluntary international accord SAICM (Strategic Approach to International Chemicals Management). The key elements of successful Chemical Leasing implementation involve process optimisation as a consequence of a more intensive cooperation between suppliers and users of chemicals, enhanced environmentally sound technology development and transfer, greening of the supply chain and capacity building, and clearly result in sound chemicals management at plant level.

Promoting Chemical Leasing based on the preventive Cleaner Production concept provides practical solutions for industry to become more efficient and at the same time reduce unnecessary hazardous chemicals consumption and protect human health and the environment. In this way, environmental and economic policy matters as well as international cooperation objectives can be addressed. In his article, T. Jakl excellently highlights Chemical Leasing and regulatory approaches in Chemical Policy.

To conclude, combining Cleaner Production and Chemical Leasing has proven to be a win-win approach for the economy and the environment. It can be applied to industries of different sizes in developing countries and countries with economies in transition and fosters sustainable industrial development and the implementation of international chemical policies. Quality assurance is an important element of successful projects and efficient tools are required to ensure smooth implementation at plant level.

With the present publication we hope to contribute to the greening of industry and inspire companies all over the world to engage in Cleaner Production and Chemical Leasing business models.

References

Jakl T, Joas R, Nolte R, Schott R, Windsperger A (2004) Chemical Leasing: an intelligent and integrated business model with a view to sustainable development in materials management. Springer, Wien

Schwager P, Moser F (2006) The application of Chemical Leasing business models in Mexico. ESPR 13(2): 131–137

Strategic Approach to International Chemicals Management – www.unep.ch/saicm

United Nations Industrial Development Organization – Cleaner Production Programme, www.unido.org/cp

United Nations Industrial Development Organization – Chemical Leasing, www.bipro.de/_clhome/index.htm

Chapter 3

The Concept of Chemical Leasing

Reinhard Joas

With the worldwide emergence of financial markets and competition becoming more and more intense, companies around the globe are continuously urged to become more efficient. This striving for increased efficiency and reduced production costs has several implications. Extensive transportation activities, low prices and high consumption rates, mass layoffs or critical working conditions in various industry sectors can be considered problematic effects, while on the other hand public welfare seems to increase continuously.

It is obvious that the striving for efficiency has been proved to be a strong driving force for changes. Consequently, this driving force fuelled by globalisation should be as well used to improve the environmental situation and a sustainable resource management. An ideal concept would lead efficiency increase to link economic and environmental interests, without endangering jobs. An ideal concept would also use efficiency increase to improve the cooperation between industrialised and developing countries. Technology transfer might be used as an element of efficiency increase, generating new markets instead of reducing jobs.

How would a concept look like that fulfils all these requirements?

One answer to this question is: Chemical Leasing (ChL).

Basic principles of ChL

Traditionally, chemicals are sold to customers, who use them to fulfil certain functions. Their suppliers have a clear economic interest in increasing the amount of chemicals sold ("The more you sell, the more you earn"). Typically their earnings increase if they sell chemicals at higher prices or to larger amounts. Higher prices, however, are difficult to be achieved due to international competition. So a main focus is set on higher sales volumes. This is in many cases related to problematic releases to the environment and to negative consequence for the future availability of resources.

Chemical Leasing (ChL) inverts a supplier's commercial interest in higher consumption of chemicals. In a ChL business model the chemical supplier is paid for the *service* provided by the chemicals and not for the *amount* of chemicals delivered. The chemical supplier becomes a service provider, and as such it is interested in keeping costs low while providing the service demanded to its customers. Reducing costs means reducing the consumption of chemicals which in a ChL business model has become an expense factor for the chemical supplier. Using its know-how regarding the substance, the chemical supplier will try to make the chemical application as efficient as possible. But efficiency not only depends on the chemical but also on the production process. An optimised production process, adjusted to the specific chemical, reduces on the one hand the amount of a chemical needed, and on the other hand additional savings like reduced energy consumption can be realised. This goes in line with the intention of the user of the chemical. Therefore both partners are motivated to find a common solution for optimising the process.

The following picture summarises main objectives and the concept of Chemical Leasing (Fig. 1).

The main players within the Chl models are typically the producer (supplier) of the chemicals and the user of the chemicals. Their functions within the business model will be described in detail in Chapter 4 and 5, where different pilot projects and related experiences will be introduced.

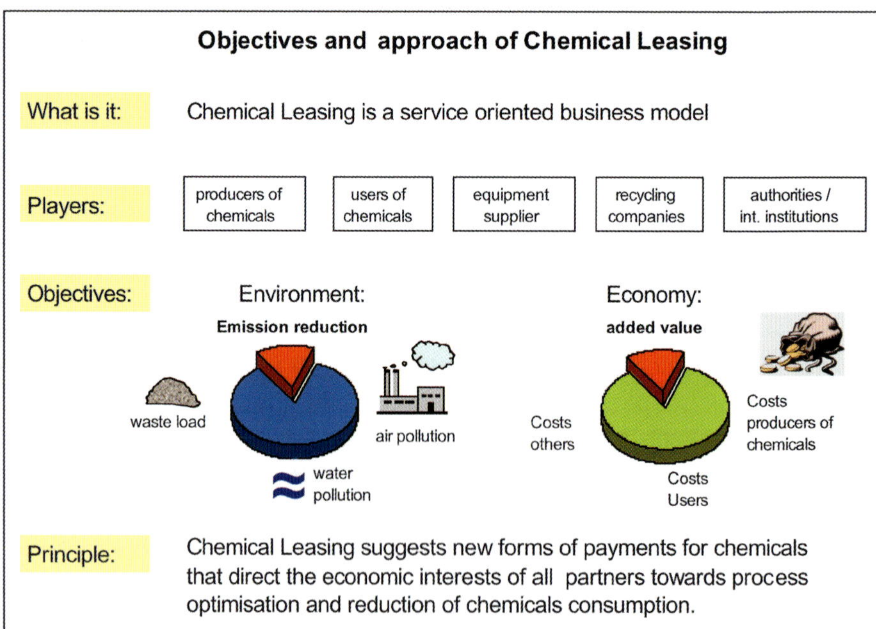

Fig. 1. Overview on objectives and concept of Chemical Leasing

The unit of payment is of high importance in Chemical Leasing. In contrast to traditional business models the basis of payment is not the amount of the chemical used (e.g., in $ per ton) but functional units like "m^2 cleaned surface".

The principle behind these units of payment leads to a "less is more" situation – this means higher earnings for the supplier within Chemical Leasing while less chemicals are provided. And it means higher earnings for the user because its costs are reduced. In Chemical Leasing models the added value is shared between the partners. Efficiency increase is the interest of both partners; both partners have a commercial interest in reducing the consumption of chemicals as both partners increase their profits.

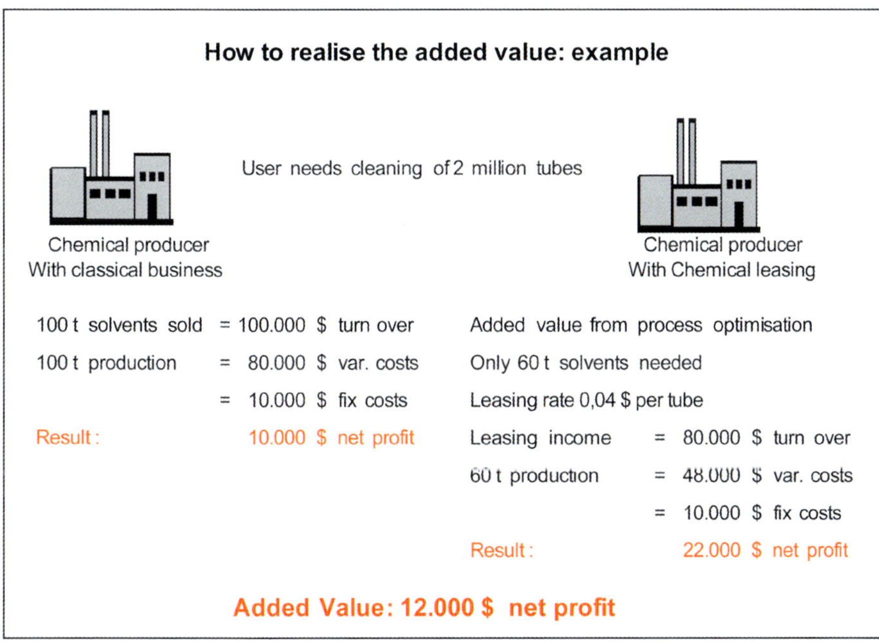

Fig. 2. Example chemical producer: Realisation of added value. In the classical business model 100 t of solvent are needed for the cleaning process. Consequently the chemical supplier sells 100 t of the solvent with an annual turn over of $ 100,000 and a resulting net profit of $ 10,000 after subtracting variable and fix costs (see How to realise added value: example)

chemical producer with classical business		chemical producer with Chemical Leasing	
$ 100,000	$ 0.04 per tube	$ 80,000	
$ 80,000		$ 48,000	
$ 10,000		$ 10,000	$ 22,000
$ 12,000			

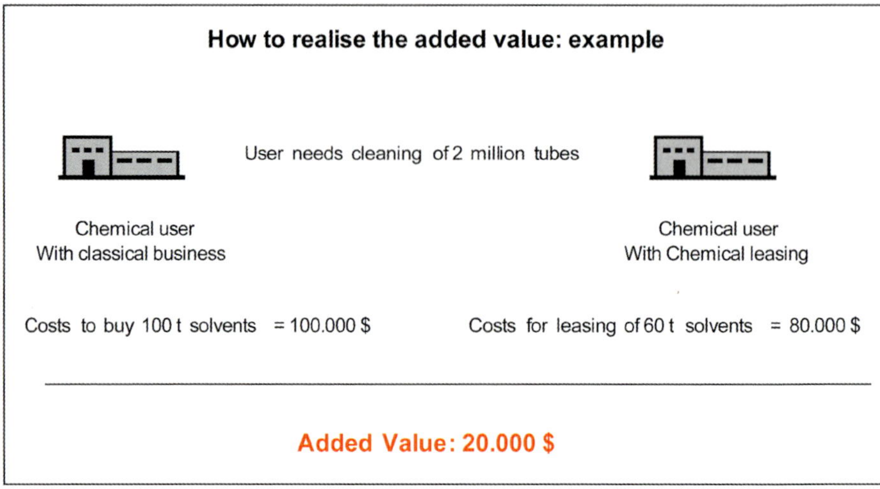

Fig. 3. Example chemical user: Realisation of added value. On the other side, the chemical user has to pay $ 20,000 less in comparison to the classical business model for the same result (see How to realise added value: expample)

chemical user with classical business
$ 100,000
$ 20,000

chemical user with Chemical Leasing
$ 80,000

The following simplified example might further explain the principle. Various aspects such as administrative efforts, improved logistics, training effects, better handling, competition and capacity impacts, etc. are not considered at that stage.

A tube producing company needs solvents to clean their products before delivery to customers. The classical model and the Chemical Leasing business model are compared with regard to the cleaning of 2 million tubes, always taking into consideration both, the supplier of solvents (chemical producer) and the tube producer (user of chemicals).

Applying the Chemical Leasing model, the supplier together with the user increase efficiency by optimising the cleaning process and by using a chemical which is optimally adjusted to the specific process and cleaning tasks. Therefore, only 60 t of solvents are necessary for the same cleaning volume. The partners agree on $ per cleaned tube as payment unit. (Remark: In practical case studies the potential reduction by process and chemical optimisation is often higher than the 40% that are applied in this model.)

Based on an agreed leasing rate of $ 0.04 per tube a turn over of $ 80,000 (leasing income) results for the chemical producer. For the production of 60 t of solvents variable costs of \$ 48,000 occur. After subtracting the variable and fix costs (\$ 10,000) from the turn over a net profit of \$ 22,000 is achieved. This means an added value of \$ 12,000 for the chemical producer by applying the Chemical Leasing model in comparison with the classical business model (Fig. 2).

This example basically demonstrates the principle of Chemical Leasing. In practice different models with different partners and characteristics exist and several further impacts have to be considered. But the principle of economic advantages for all parties involved combined with environmental benefits due to reduced consumption of chemicals or other resources (resulting in reduced emissions to the environment) is valid for all types of Chemical Leasing.

Advantages of ChL

Cost savings achieved by means of reduced chemical consumption are ultimately shared among the chemical supplier and its customer. On the one hand the chemical supplier has less production costs and on the other hand its customer has to pay a lower price to obtain the benefit desired than if it bought the chemicals and applied them traditionally. Applying this business model is thus of economic advantage for both partners. Chemical Leasing is additionally advantageous for chemical suppliers since they will be able to strengthen customer relationships. For customers on the other hand it is advantageous to concentrate on their core business and to cede responsibility for the management of chemicals. Both partners clearly benefit by a more innovative and positive public image.

In addition to the economic advantages ChL also has obvious environmental advantages. Process optimisation not only leads to a reduced chemical consumption but very often also to a reduction in the consumption of other resources like energy or water. As a result the waste load as well as air and water pollution will decrease, reducing the total environmental impact of the production process.

In the case studies in Chapters 4 and 5 possible economic as well as environmental benefits are described in more detail.

Actors of ChL and international approach

The key player in any ChL business model is the supplier of chemicals. And it is essential that the supplier has know-how about the chemicals and their efficient application. In many cases the supplier will also be the producer, but also a trading company with sufficient know-how can be a proper supplier. The primary partner of the chemical supplier is its customer, who benefits from the chemical being used in its production process. Optionally additional partners can be involved in the ChL business model. The manufacturer of the equipment in which the chemical is used can also contribute valuable knowledge on how to optimise the chemical application and help increasing efficiency and further reducing the chemical consumption. In some cases the equipment producer might even play the most important role in the Chemical Leasing business model (see Chapter 4.2).

Specialised recycling companies might also be involved in the business model as further partners if they can help to improve the efficiency of use.

Besides these actors from industry, a series of stakeholders can play a role in ChL. To establish the business model it is useful to involve a party who can act as a mediator between the different partners and raise confidence among them. This task can be addressed by private consultants, but in particular institutions specialised in providing advice to industry, such as the UNIDO Cleaner Production Centres, can fulfil such a role (for further details see Chapter 2). To strengthen confidence in the ChL business model as such, it is important to involve an organisation which can provide independent quality assurance and thus guarantee that the main principles of ChL (economic and environmental benefits) are kept (for further details see also Chapter 2). Ultimately, a business model enjoying general confidence like ChL can be used by national governments as a political tool to improve risk management practices and extend the chemical producers' responsibility on a voluntary basis (for further details see Chapter 13).

Definition of ChL

For the dissemination and the establishment of the ChL business model in different industries and different countries a clear definition of ChL is essential. As the United Nations Industrial Development Organisation (UNIDO) plays a leading role in the development and implementation of ChL around the world, it was UNIDO that defined – together with an international working group – the term of Chemical Leasing (see Chapter 2).

The definition starts by pointing out the service-oriented character of the ChL business model. This is linked to the innovative shift in payment basis from product based payments to payments based on functional units. The extended producer responsibility referred to in the definition results from the ownership of the chemical not being transferred from the supplier to its customer. Remaining the owner of the chemical, the supplier is responsible for the application as well as for the disposal or recycling of the chemical. Increasing the efficient use of chemicals while reducing the risks, as stated in the definition, are the targets of any process optimisation under ChL. To optimise a production process involving the application of a chemical, the chemical and the equipment used have to perfectly fit together. For this purpose the chemicals applied can be replaced by chemicals more adapted to the equipment and process in question. Such customised chemicals are developed best combining the know-how of the chemical producer and the manufacturer of the equipment. The definition concludes by naming the preconditions for the successful implementation of ChL business models.

ChL going global

At the 2002 Johannesburg World Summit on Sustainable Development the international community agreed on the goal of ensuring that, by the year 2020, chemicals will be produced and used in ways that minimise significant adverse impacts on human health and the environment.

As a consequence the United Nations Environmental Programme (UNEP) decided to instigate the development of a Strategic Approach to International Chemicals Management (SAICM) in 2003. For this purpose a multi-stakeholder and multi-sectoral Preparatory Committee should analyse the chemicals management practices around the world, identify the deficits and determine the prerequisites for international action on chemical hazards. In 2006, at the International Conference on Chemicals Management (ICCM) in Dubai, the participating countries finally resolved upon SAICM by signing the Dubai Declaration on International Chemicals Management. The signing countries thereby committed themselves to promote the sound management of chemicals and hazardous wastes at all levels. An effective means of translating this idea into public policy is the implementation of ChL. The concept of ChL is particularly in line with SAICM, because both aim at a sound management of chemicals throughout their life-cycle, and a responsible way of use that leads to the minimisation of adverse effects on human health and the environment. SAICM wants to encourage support of developing countries in strengthening their capacity for the sound management of chemicals. If applied in a transnational context, ChL can contribute precisely to this objective and to technology transfer from industrialised to developing countries in general. This conclusion was also drawn at the UNIDO-Austria side event at the ICCM in Dubai where the concept of Chemical Leasing was introduced to the international community. More than 100 representatives from governmental institutions, industries, IGOs and NGOs participated in this side event.

Different ChL models

Depending on the general conditions – like number of involved partners, offered service – three main models of Chemical Leasing can be distinguished (see Fig. 4).

Model A is a very simple model which includes only two partners (chemical producer and chemical user). This model is often mixed up with the commonly established "recycling system". Although at first glance the principle is the same – the producer of the chemical collects the used chemical and delivers a "new" one – there is one major difference within the „recycling system": The user pays for the amount of chemical whereas according to the Chemical Leasing model the user pays for the function of the chemical (e.g., m^3 cleaned water instead of activated carbon).

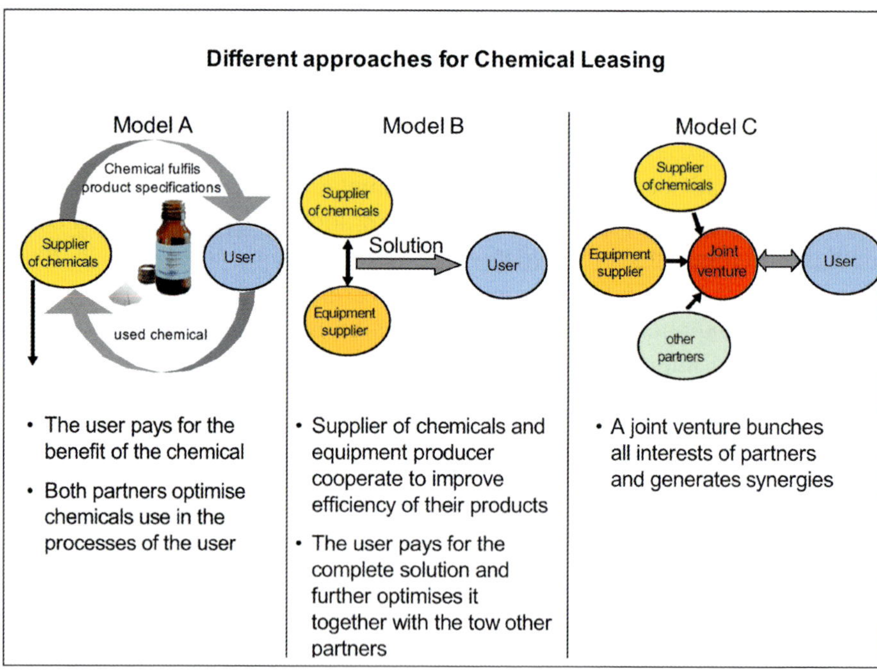

Fig. 4. Models of Chemical Leasing.
Models of Chemical Leasing Other partners
Both partners optimise the use of chemicals in the user's processes
… co-operate to improve the efficiency of their products
… together with the two other partners
A joint venture combines all interests …

Model B is more complex and besides user and producer of the chemical also the equipment supplier with its specific process know-how is involved in the process. In this model the chemical producer together with the equipment supplier offer a solution that has been improved with regard to efficiency of chemicals and energy use. The user pays for the benefit of the complete solution and further improves efficiency of the offered solution related to its process.

The most complex model is Model C which involves several partners that all follow the principles of Chemical Leasing and combine their interests in a joint venture.

Problems to overcome by ChL

Even though ChL is designed to offer advantages to the involved parties, until now ChL is no self-running process. An initiator and some „catalysts" are necessary to start the „reaction". Possible initiators are international organisations like UNIDO with a worldwide network of National Cleaner Production Centers but also companies which already successfully apply ChL might be a driving force for this business model.

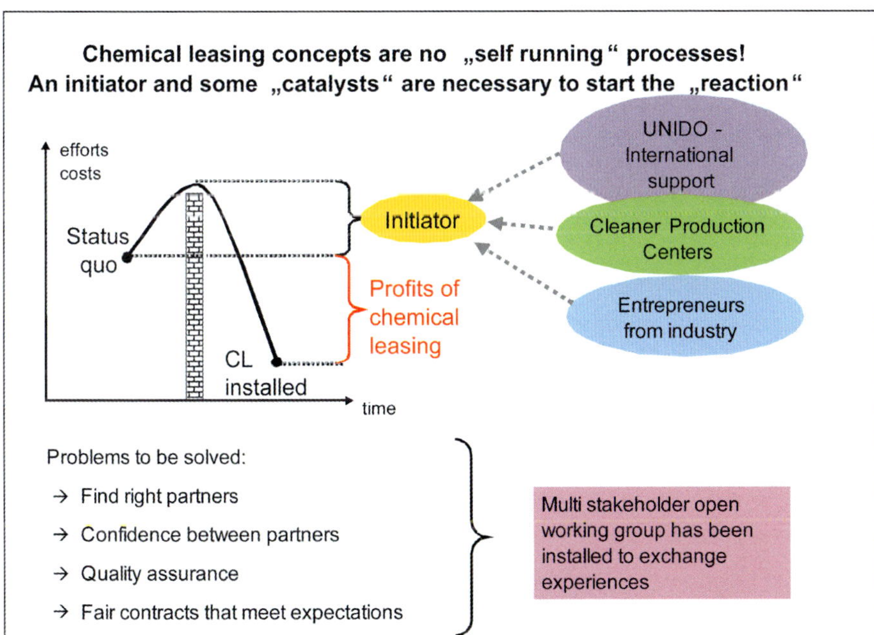

Fig. 5. Main barriers for the implementation of Chemical Leasing

As shown in Fig. 5, the main problems which have to be overcome for the implementation of ChL are the following:

- Find right partners;
- Confidence between partners;
- Quality assurance;
- Fair contracts that meet expectations.

Potentials for expansion of ChL and Outlook

Since chemical products provide a broad variety of services such as "cleaning", "coating", "colouring" and "greasing" the ChL model is applicable in a multitude of industry sectors. In addition ChL is applicable to large companies just as to small and medium enterprises. Based on the experiences obtained to date, ChL business models have the greatest success when applied to processes that are not yet optimised with regard to chemicals consumption.

The concept of Chemical Leasing can also be extended to other fields. An example can be seen in the improvement of energy use by an intensified co-operation of supplier and user. Again the result will be a win-win situation for both, environment and economic profits of supplier and user.

All parties involved in the implementation of the concept are aware that several difficulties exist related to first pilot projects. But the number of success stories increases continuously and it is expected that Chemical Leasing will be seen as one of the major answers to the challenges of globalisation, sustainable development and economic improvements.

Chapter 4

Chemical Leasing in Austria – Case Studies

4.1 Enhancing the Acceptance of Chemical Leasing

Christian Plas

The problem explained

The concept of Chemical Leasing comprises different business models that all have the goal of abandoning classical customer-supplier relations for integrated approaches where different levels of responsibility are transferred from the consumer business to the chemical supplier.

It is an essential feature of Chemical Leasing that the quantity of a specific substance no longer defines the basis for business, which is instead represented by the function a certain substance can produce (e.g. as cleaning, dissolving, or reactive agent). A further feature is that the ownership of chemicals is no longer passed on, but that substances remain in the possession of the supplier (i.e., no transfer of ownership).

By this means, none of the parties involved has an interest in the quantity used of a given chemical. On an environmental-political level, a better conservation and thus resulting in increased safety is to be expected.

The business models for Chemical Leasing can offer the user several advantages:

- Cost advantages,
- Avoidance of inefficient capacities,
- Optimised chemical usage,
- Increased safety, and
- Accessibility to "chemical know-how".

Nevertheless, a number of reservations exist:

- Dependency caused by a too narrow customer-supplier relationship,
- Reduced flexibility,
- Know-how leakage to competitors,
- Uncertainty with compliance to technical specifications, and
- Technical and logistical problems with delivery and removal.

Despite these impeding factors, business companies interested in Chemical Leasing are asked to recognise potential usage, work out its specific advantages, and develop a viable business model with partners.

Business companies generally fear the potential loss of influence and possible procedural impasses that they associate with changes in the organisational structure or production processes. Arguments for such a change should thus be fairly concrete and preferably quantifiable to facilitate their realisation.

In practice, it proves to be difficult to abandon well-functioning production methods and operations and consider improvements. This report attempts to expand the strategies for changes in order to facilitate their implementation, and hopefully increase their appeal as well.

Developing a method

Chemical Leasing can begin with traditional check-lists and questionnaires to find a theoretical basis. This aims at awakening a willingness for change by way of creatively describing the desired solution.

Only in the next step technical and organisational aspects should be considered. They should be described as independently as possible leaving aside the future concept, in order to avoid preconceptions that could threaten the changes altogether. In this way, acceptable and realisable Chemical Leasing projects will be sought for the respective business (Fig. 1).

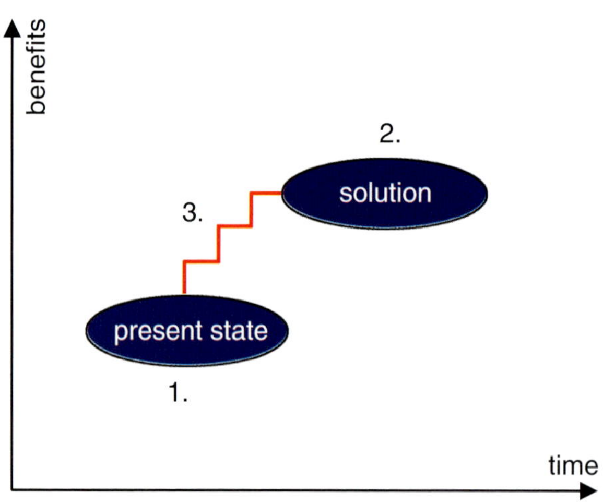

Figure 1

The illustration schematically describes the suggested strategy:

1. The current state is first described and evaluated (with regard to advantages and disadvantages for the respective business, for its employees, the environment and other stakeholders that might be involved).
2. The evaluation of Step 1 serves as a basis for the question "How could operations be improved?" Step 2, "Description of the Solution" re-counts the development of a desired future state for the repective business. In each process, all persons directly involved find aspects needing improvement. Ideal solutions, free of feasibility constraints, will be described showing how optimised operations could run eliminating the current weak points.
3. Only in the last step feasibility will be checked, with a clear solution in mind. General conditions will be determined and evaluated whether they are available or need to be created. This partial step will be called "Finding possible solutions", the result of which is a list of tasks needed for determing the feasibility of a new business model for Chemical Leasing.

In the following, the partial steps will be explored by way of an example:

Description and evaluation of the current state

In case Chemical Leasing should only partially be applied, operations and facilities involved should first be exactly defined. These will then be described, as well as their associated functions and tasks.

Considering outputs and inputs, the conceivable application of Chemical Leasing will be identified for the relevant areas of operation.

Emissions (waste in solid, gas, and water form) with appreciable quantities of reclaimable or reusable chemicals, and uses of chemicals with a performance efficiency above 80% are predestined for Chemical Leasing employment.

The current state of the case study illustrated (Table 2) shows that technologies for reclamation (condensation, distillation, metal reclamation) or recycling (electrolyte production) are for the most part available for process emissions. Furthermore, chemicals with poor efficiency are currently used in the area of electroplating.

The use of perchloroethylene as a cleaning agent, the use of chemicals containing Ni, Cu and Sn for surface treatments, and the use of activated charcoal for waste water purification are preferred areas for Chemical Leasing implementation.

Table 1. Example of an evaluation table for identifying potential optimisation

Potential	Factor	Costs for purchase and disposal of wastage	Admin. expenditure for purchase and disposal	Expenses for storage of chemicals	Expenses for facility education and training	Expenses for facility maintenance and repair	Hazardous qualities of chemicals	Frequency of handling	Environmental hazards of chemicals	Problems of waste management
None or negligent	1									
Low	20									
Average	50									
High	70									
Very high	80									

Table 2. Evaluation results in a case study

										Potential
Potential	Cost for purchase and disposal of wastage	Admin expenditure for purchase and disposal	Expenses for storage of chemicals	Expenses for facility education and training	Expenses for facility maintenance and repair	Hazardous qualities of chemicals	Frequency of handling	Environmental hazards of chemicals	Problems of waste management	Weighted total
Weighting	**50%**	**5%**	**5%**	**5%**	**5%**	**10%**	**10%**	**5%**	**5%**	**100%**
Perchlor-ethylene	1	20	70	1	1	70	70	100	50	**26,6**
Metal salts	70	70	70	1	20	50	100	50	70	**64,1**
Activated charcoal	1	50	20	1	20	20	70	1	50	**16,6**

Table 3. Assignment of individual evaluations and priority numbers to the potentials

Single evaluation	Total potential		Colour scale	
1–50	0–29	light yellow		little potential
50–70	30–60	orange		average potential
70–100	60–90	red		big potential

Proceeding in line with the "Manual for Chemical Users", an evaluation table of potential optimisation is developed, outlining a process to be further defined. In addition to the criteria "costs", "risk", "procedure", the table includes "environmental relevance". Before evaluation, the table's categories ("marginal" to "very high") should be adapted by operational units to fit their specific requirements (Table 1).

The weighting helps to determine at which percentage a criteria should be calculated in the total evaluation. The sum of all weighted evaluations results in a priority number which expresses how worthwhile an alternative business model would be (Table 2).

In the case study, the criteria "costs" are calculated into the total evaluation with a factor of 50%.

Job safety is also of significance, considering criteria like "Chemicals hazardous for people" and "How often must it be handled?". The remaining criteria are equally weighted. Results show one usage with an average potential, and two usages with little potential for improvement with Chemical Leasing (Table 3).

The result of this evaluation is a list of conceivable applications of Chemical Leasing, evaluated according to the potential improvement for the user. The evaluation allows identification of the chemical usages in which Chemical Leasing would result in the highest benefit.

Work will be continued with the three case usages showing the highest potential.

By this step, the applied evaluation process (value benefit analysis) is used to help the business establish clarity and objectivity in assessing the situation. Colour coding helps to better understand the potential improvement. "Traffic light illustrations" distinctly show the complexity of issues.

Finding possible solutions and decriptions of solutions

At the beginning of the creative processes for finding possible solutions, it is important to be aware about how creative thinking happens. Simple exercises tailored to group size and situation (e.g., 9-Point Exercise, as described in the book "Ganz im Gegenteil", Varga von Kibed[2]), can initiate the willingness to change and allow positive creative thinking.

In a group exercise, the current situation of relevant usages will be examined from different perspectives. Each group is assigned a specific role and works out solutions from this perspective (suggestions for improvement).

Examples of roles and questions could be:

- How would a family of four solve this problem?
- How would a safety engineer handle it?
- Which possible solutions would occur to Albert Einstein?

Viewing the questions from a distance and from a completely new perspective, solutions will be found that never would occur if conventional thinking is applied. For this reason, operational feasibility is ignored in this step. Yet, depending on the specific role, comprehensive solutions will be sought (even outside the boundaries of the actual operation). (Everything is possible; it is just a question of general conditions.) At the end of the group exercise three different solutions for the three case studies are read to be presented to the plenary session.

In a general discussion, it will first be attempted to define ideas exactly ("bring them down to earth").

For each solution, the question will be asked: "What is needed for realisation?". By this means, the necessary "hard" and "soft" pre-conditions will be determined (e.g. new partners, specialists, choice of business management, different company culture, …) In a further step, these pre-conditions defined are further described by a timeframe: "When will it be possible?" During discussion, tasks necessary to clarify pre-conditions will be continually noted. Wherever there are potential improvements described, which cannot be optimised in-house, links to new business models like Chemical Leasing are established.

Conclusions

It is often difficult to understand from an outside perspective why optimisation solutions are not implemented. Is is specifically such changes that demand a re-thinking in business operation, for example no longer possessing the chemicals employed, which are sometimes met with great scepticism.

As with many other operational problems, the answer to allowing further development does not lie in improving technical productivity or enhancing amortisation of investments, but rather in stimulating different ways of thinking.

The lack of clarity of new processes, new dependencies, endangered process safety, less control options, and increased liabilities act as stumbling blocks that inhibit fundamental reform.

In order to set the ball of the reform process rolling several methods described here can be applied:

- Improved and more transparent representation of the potential change.
- Search for optimised production circumstances, without being too confined by existing general conditions, and
- Working out solutions including changed pre-conditions.

The use of creative thinking methodology is strongly recommended when dealing with new terms of production. They enable a more comprehensive and better defined result than conventional methods would in the same time.

That means: Dare to do it! Chances will be good that the increased use of Chemical Leasing in industry and trade will make a big step towards improved resource management, job safety, and protection of both neighbourhood and environment – and be economical at the same time.

4.2 A Best Practice Example of Chemical Leasing in Metal Cleaning in the Automotive Industry

Report by an Austrian company

Horst Erbel

The purpose of Chemical Leasing

Chemical Leasing, in its broadest sense, reaches far beyond the use of chemicals: According to this concept, the operating process, as a whole, is being leased. Thus, in addition to chemicals, the technical facilities, all services necessary for smooth functioning and, if applicable, also human resources are provided. This helps to dramatically improve the reliability of the complete production process.

In our case, Chemical Leasing is covering the following aspects of the production process:

- Chemicals,
- Processing,
- Technical facilities,
- Services (for facilities and chemicals).

As a consequence, there is a common intention of all parties involved in this model to reduce the consumption of chemicals and energy. In addition they all are also interested in a high reliability of the facilities and in minimising services to be provided – for the benefit of all partners involved and the environment as well.

Chemical Leasing and PERO AG

The idea of Chemical Leasing was developed by the Austrian Federal Ministry of Agriculture, Forestry, Environment and Water Management (BMLFUW). PERO AG was fascinated by this idea and decided to support its realisation.

PERO is a manufacturer of high quality metal cleaning machines and has always followed the goal of reducing the amount of chemicals consumed by their machines. For example, 50 years ago the introduction of electric refrigeration reduced the amount of chemicals consumed by the cleaning machines by more then 50%; 25 years ago, PERO AG developed the first encapsulated machines; 15 years ago, the limiting values provided by German Immission

Protection Act were based on PERO machines which were state-of-the-art; seven years ago, they developed the first machines functioning without the error-prone electric refrigeration.

In order to realise the idea of the Austrian Federal Ministry, PERO AG had to find a suitable partner with adequate know-how about chemicals. They selected SAFECHEM Europe GmbH, Düsseldorf, a subsidiary of The Dow Chemical Company. The main decisive factor for their collaboration was that SAFECHEM was able to provide the chemical know-how as well as logistics required.

The contact details of SAFECHEM Europe GmbH are the following: SAFECHEM Europe Gmbh, Georg-Glock-Straße 3, 40474 Düsseldorf, Germany.

Contact person for Chemical Leasing: Mr. Steffen Säcker

Webpage: *www.safechem-europe.com*

Foundation of PERO Innovative Services GmbH

For the realisation of this project, an appropriate company was founded: PERO Innovative Services GmbH, Bundesstraße 47, 8160 Weiz, Austria.

Webpage: *www.pero-innovative.at*

It was agreed on the following task sharing:

PERO

- provides machines that are best qualified for the model,
- provides rooms and materials logistics,
- provides the necessary human resources.

SAFECHEM

- provides the solvent for the cleaning process,
- monitors the quality of solvents,
- provides the waste management of the used chemicals (waste oils etc.),
- provides stabilisers.

Remark: Providing rooms, staff and logistic support is actually not part of Chemical Leasing. But by offering these additional services, PERO Innovative Service has facilitated to implement this business model for the first time world-wide.

The company Automobiltechnik Blau, a branch of MAGNA STEYR Fuel Systems Ges.m.b.H. in 8160 Weiz/Preding, Austria (*www.blau.co.at; www.magnasteyr.com*) was identified as first client for the new business model.

Automobiltechnik Blau's core competence is the forming of metal parts. Main products are filling systems like caps of fuel tanks and oil tanks. Their clients are car manufacturers all over the world. Automobiltechnik Blau has been happy to find a competent service provider who is capable to perfectly clean these parts.

It is essential for achieving best results economically and ecologically that the lessor has freedom of choice how to solve the task best. This leads to a win-win situation for both economy and ecology.

The lessor has the best knowledge about possible techniques. Its high competence comprises know-how in the fields of process technologies, facility construction, properties of chemicals and handling of chemicals.

Within the lessor's company, partly contrary economical targets are combined:

- cost efficient machine,
- low consumption of energy,
- low usage of chemicals.

Thus, it is perfectly qualified to find the best possible solution for the task and to guarantee its implementation in practice.

Practical experience

A long-term Chemical Leasing contract was signed by PERO Innovative Services (as provider of Chemical Leasing services) and Automobiltechnik Blau (as client). The client pays per number of cleaned parts. A precise quality specification is defined for each individual part.

It is the model's main advantage that the user Automobiltechnik Blau (MAGNA STEYR Fuel Systems GmbH) can carry on focussing on its core competencies. The company needs no (additional) expert who understands the functioning of the technical equipment, the chemistry and the physics of the cleaning process and who keeps himself or herself up-to-date, even regarding environmental legislation.

A misinvestment (in inappropriate cleaning equipment) is prevented. The company avoids the expenditure of capital while it has high security of calculated costs. Any surprises regarding costs, like unexpected high consumption of chemicals, unexpected service requirements, etc., can be ruled out.

Of course it is a pre-condition for the user to find a reliable and competent partner who does not "throw in the towel" if problems arise. This has been guaranteed by the foundation of PERO Innovative Services.

After the foundation of PERO Innovative Services, which was accomplished in a surprisingly uncomplicated manner, the personnel required was selected and hired. The search for premises was managed quickly with the help of PERO Innovative's future client Automobiltechnik Blau.

Being both service provider and manufacturer of the equipment, PERO was not exposed to the usual cost pressure often resulting in purchasing cheaper machines while neglecting any higher cost-in-use. Of course a high-performance PERO cleaning machine was bought to run the core process.

PERO Innovative Services has developed the equipment and the peripheral devices in such a way that they optimally adapt to the desired cleaning process and minimise the costs during the complete life cycle of the equipment (according to LCC – Life Cycle Costing, formerly accurately related to as TCO – Total Costs of Ownership, since traditional accounting systems in most cases regard a period which is too short, as the costs as well as the benefits of the system incur over a long time utilisation of the equipment).

As provider of Chemical Leasing services PERO Innovative Services was able to include in the project all peripheral devices necessary for the supply, reconditioning, recovery and maintenance of the chemicals. The equipment was engineered based on the principle of cost reduction with regard to chemicals consumption, energy consumption, supply of spare parts and service. The solvent and the stabilisers used were adapted to the specific cleaning needs of Automobiltechnik Blau.

Since both, the equipment manufacturer PERO and the chemicals supplier SAFECHEM, are involved in production operations, the life expectancy of the equipment, as well as the maintenance of the solvent, can be optimised.

Altogether process costs are reduced for the benefit of the user and all other parties involved.

Intentions of the chemicals supplier have changed. It is no longer interested in achieving a large turnover of chemicals through increased sales. It is rather interested in maintaining the chemicals in use as long as possible. Recycling of chemicals instead of expensive disposal also helps the chemicals producer to save costs.

The economic success of the chemicals supplier is increasingly generated by the services provided by its chemicals. Responsible chemicals management leads to reduced sales of chemicals and subsequently to reduced revenues of the chemicals supplier. Yet a reduction of the amount of chemicals used moves the chemical product as such out of focus while the service provided, in connection with the chemical, becomes increasingly important. This includes the environmentally safe supply of chemicals, internal recycling, increase of lifetime by means of stabilisers, analyses of used product and recovery of resources.

Experiences in the pilot project

The pilot project has now been running for two years. In the beginning mutual trust between the service provider and the service receiver was of utmost importance. Especially the service provider, who supplies the equipment, has to make large investments in order to gain the confidence of his future customer. On the other hand, the service provider has to trust its client to continue to be successful also in future. But also the service receiver needs confidence in its future service provider, since it will not be able to satisfy its own customers without a perfect service provided. Proper contracts are the solid base for a future partnership. It takes reliable persons and machines, however, to accomplish a successful cooperation.

As mentioned above PERO Innovative Services rather focussed its attention in the reduction of running costs than in keeping machine investments low, in order to gain experience.

From an economic point of view, a machine, like the one which is currently used, would never be sold on a free market, where a client compares prices of cleaning equipments of different manufacturers. Very important and high-performing – and therefore expensive – components have been incorporated for reasons of reduced emission and energy consumption. By these means environmental standards are exceeded, a fact which is not rewarded in free market competition.

Within this project of the BMLFUW two other PERO customers with similar applications agreed to disclose their operating costs. PERO Innovative Services wants to take this opportunity to thank them for their support.

Significant cost reduction could be achieved:

1. Energy: reduced by 50.1%;
2. Spare parts and services: reduced by 66.4%;
3. Solvents: reduced by 71.7%;
4. Stabilisers: reduced by 76.9% and 55%, respectively.

In total the following services were provided from May 2005 to March 2007 (23 months):

Operating hours	5,890 hours
Charges cleaned	39,740 charges
Tonnage cleaned	approx. 1,495 tonnes of parts
Solvent for recycling	626 kg
Net solvent consumption	approx. 300 kg
Total stabiliser consumption	22.5 L
Energy consumption	245,226 kWh

However, in the first year these savings were not sufficient to amortise completely the costs of the foundation of the company and the depreciation of the machine.

Nevertheless, credit can be taken for the successful savings in energy, solvent and stabiliser from an ecological point of view, which adds to the general acceptance of this high performing technology. This had been the original target of the BMLFUW.

For the second year, further cost savings and a moderate price increase accepted by Automobiltechnik Blau, will ensure the company a modest profit which will be reinvested into further marketing of this new and sustainable concept.

Due to the ongoing growth of Automobiltechnik Blau and the competitive model a further investment into the Chemical Leasing model is expected for the next future.

Please note: We should not underestimate the fact that the personnel of PERO Innovative Services is one of the key success factors: They exclusively control the cleaning process and the quality. Their only task is to run and maintain the machine. Their experience and skills in stabilising ensure the successful operation of the process as a whole.

All of the staff were trained accordingly. Daily work has further added to their expertise in cleaning. They are able to run the machine proficiently and to immediately notice any malfunction. Such expertise is needed to successfully deliver high quality services which can be charged to the customer.

Conclusions

The Chemical Leasing model has proven its ability to run in real operation for a sufficient period of time. Therefore PERO decided to offer this new concept also to a broader range of customers. PERO Innovative Services would be happy to work out such a concept in cooperation with other operators of cleaning machines. This would not only be a win-win situation for all parties involved regarding cost accounting but also for our environment.

Fig. 1. PERO Cleaning Machine (from left) Thomas Jakl and Horst Erbel

Fig. 2. PERO Cleaning Machine

Questionnaire

A Best Practice Example of Chemical Leasing in Metal Cleaning in the Automotive Industry

Report by an Austrian company

PERO AG

			① LOW	②	③	④ HIGH
1	What were the key benefits expected leading to your participation in the project? **Influence on the application of the procedural plant which increases live expectancy; pressure on the market.**					
2	Were these goals achieved?		YES X		NO O	
	If not, why:					
3	Was the implementation of the new business model challenging in general?		①	②	③	❹
3	a	Difficulties during/with the implementation? **Convincing the user and winning people's confidence.**				
4	Is the new business model competitive		①	②	❸	④
5	Are you successful with the new business model?		①	②	③	❹
6	Could you use the experiences also for other products/offers?		①	②	❸	④
7	How many customers have accepted the new business model till now?		1			
8	Are these small (to 50 employees), medium-sized (51 to 250 employees) or big enterprises (more as 250 employees)?		medium-sized			
9	Is the application of chemicals a core competence of these enterprises?		YES O		NO X	
10	Were other companies involved in the implementation of this model (providers of equipment, disposal/recycling companies, etc.)?		YES O		NO O	
	If yes, please specify: **SAFECHEM**					
11	Characteristics of involved chemicals:					
		Value	❶	②	③	④
		Recyclability	①	②	③	❹
		Hazardousness	①	②	❸	④
12	Consequences of the model:					
		Product innovation	①	②	③	❹

		Earnings	①	②	❸	④
		Cost	①	❷	③	④
		Sales volume: **€ 246.716,00/year**	①	②	③	④
		Customer relationship	①	②	③	❹
		Competition	①	②	③	❹
		New customers: **Name "Chemical Leasing" is deceptive and not suitable**	①	②	③	④
		Mass of raw material/new product (also in %) **% to which basis?**	❶	②	③	④

13	Is the model also suitable for very small enterprises (5 to 20 employees) or small product quantities?	YES O	NO **X**
	Reasons: **Implementation needs to work with full capacity to charge off!**		
14	Motives for the introduction of the new business model? **Influence on the application of the procedural plant which increases live expectancy; pressure on the market. The user does not need expert knowledge anymore and can concentrate on its core competence.**		
15	What are the main advantages/disadvantages of this model as compared to the traditional sales concept? **Influence on the application of the procedural plant which increases live expectancy; pressure at the market. No initial investment at user's side and a most reliable process without requiring an expert. The supplier satisfies all legal formalities with the authorities referring to the laws and environmental regulations.**		
16	Which charging base (unit of payment) do you use?	**payment per cleaned part**	
17	Has this charging base any disadvantage from your point of view?	YES O	NO **X**
	If yes, please specify:		
18	What quantity of a new product could be saved? kg, l, %, etc.	**120 kg/year**	
19	What quantity of hazardous wastes could be reduced? kg, l, %, etc.	**no hazardous wastes**	
20	What was the decrease of emissions to the environment? kg, l, %, etc.	**97%**	
21	Would you implement the new business model once again?	YES **X**	NO O
22	Will you use the business model also for other products? **Unknown!**	YES O	NO O
23	Will this model foster the implementation of REACH?	YES **X**	NO O
	Reasons: **REACH becomes redundant with this business model.**		
24	What you would change, complete or develop at this model? **We suggest to change the name "Chemical Leasing", e.g. to: "process rental".**		

4.3 Chemical Leasing in the Field of Paint Stripping

Walter Beyer

Case study information:

1. Involved partners and country information

- Company name — Mepla-Alfit
- Field of activity — Production of hinge, drawer and slide systems
- Role in the project — Service recipient
- Location — Austria, 6840 Götzis, Sennemahd 10
- Contact person — Stefan Schmid
- Website — www.alfit.com
- Information about the industrial sector and conditions in the country

The company needs special hooks for the process of painting the different parts. An organic solvent mixture is used for internal de-lacquering.

During the last years Tiefenbacher GmbH has improved its facilities and finally implemented a de-lacquering plant including aspiration and bath reconditioning. In spite of the reconditioning the de-lacquering baths have to be exchanged on a regular basis. The de-lacquering baths and the separated paint residues (filter cakes with high residual solvent content) have to be disposed as hazardous wastes.

A part of the solvents transported into the water by the necessary –final cleaning of the hooks, the waste water is fed into the internal water cleaning system. Outsourcing the de-lacquering process is not feasible as the daily number of hooks for de-lacquering is up to 8,000 pieces. When outsourcing a multiple amount of hooks would be necessary and the transport effort would increase considerably as well.

- Company name — Tiefenbacher GmbH
- Field of activity — de-lacquering
- Role in the project — Service provider
- Location — Austria, 4482 Ennsdorf, Industrieparkstraße 3
- Contact person — Alfred Tiefenbacher
- Website — www.tibagmbh.at
- Information about the industrial sector and conditions in the country

Tiefenbacher GmbH is a chemical-technical company that delacquers metal, plastic and wooden surfaces for its clients using mechanical, thermal and chemical processes.

The solvent mixture used for the de-lacquering is prepared by Tiefenbacher themselves. Optimisation of workflow and improvement of the solvent mixture by integrated bath reconditioning and solvent treatment led to a reduction of the feedstock by 50%.

The composition of the solvent mixture was also improved and optimised for the use as de-lacquering agent.

An additional benefit of the internal conditioning is the avoidance of hazardous wastes and significant (≤ 50%) reduction of the waste volume.

Besides the core business of paint removal they also started to supply solvent mixtures for de-lacquering to other companies.

- Company name BEYER Umwelt + Kommunikation
- Field of activity Management consulting
- Role in the project Consultant
- Location Austria, 1060 Vienna, Gumpendorferstraße 92
- Contact person Walter Beyer
- Website www.beyer.at
- Information about the industrial sector and conditions in the country

BEYER Umwelt + Kommunikation has been a consultant for both companies in the introduction and implementation of the project.

General Information about the industrial sector in Austria

In the following the economic situation in Austria is presented according to the survey "Struktur und Entwicklung der Industrie Österreichs" by the Industriewissenschaftliches Institut, study no. 126, published by the Austrian Economic Chamber representing the Austrian industrial sector and the Federal Ministry of Economics and Labour of the Republic of Austria, March 2006.

The comparison of the Austrian industrial sector with the Austrian trade sector shows that industrial enterprises amount to 2% of the total of 242,886 companies (Source: Sonderauswertung der Leistungs- und Strukturstatistik 2003 [special evaluation of performance and structural statistics]). In 2003 the 4,800 industrial enterprises created more than one third of the production value of the total industry (35.6% of EUR 264.5 bio) and more than a quarter of the current gross revenue (26.3% of EUR 120.6 bio)

The three biggest industrial segments in 2003, taking into account the current revenue of the general industrial sector, are:

- Machinery and metal goods industry (6.01%);
- Electro and electronic industry (3.26%);
- Chemical industry (3.16%).

The total value created by these enterprises amount to 12.4% of the total industry.

The size pattern of those enterprises in the industry deviates significantly from that of the business enterprises. While in Austria's industrial sector 45.5% of the companies are counted as very small ones (maximum 9 employees), in the business enterprise – related segment that number is almost double (86.4%). Roughly one quarter of the industrial companies are small ones (10 to 49 employees) and one fifth are medium sized ones (50 to 249 employees).

The Austrian industrial landscape 2003 is characterised by small and medium sized enterprises (SME). A special survey of the Leistungs- und Strukturstatistik (performance and structural statistics) 2003 showed that 91.3% of the enterprises employed less than 250 people.

2. Description of processes and services supplied according to the Chemical Leasing model

Since outsourcing de-lacquering is impossible due to organisational and economic reasons the know-how of a de-lacquering company was to be implemented in the company by way of Chemical Leasing. Outsourcing de-lacquering would also lead to an overall higher environmental burden due to transport requirements and a higher use of resources due to a higher demand of hooks.

According to the Chemical Leasing model no product is purchased. Instead the total management of the material flow is to be accomplished by the provider.

Like in outsourcing de-lacquering the service of de-lacquering itself is also included in Chemical Leasing. The total chemicals management needed is outsourced to the de-lacquering company and supplier of the solvent.

By implementing this model the lifetime of the solvent mixture is expected to be longer, hazardous waste will be avoided and the application of primary solvents should be reduced.

The service recipient needs special hooks for manipulating the parts. In order to guarantee a high quality of the painting surface it is necessary to delacquer the hooks after one loop. The defined de-lacquering time is also an important factor.

At the beginning of the implementation of the new business model the service recipient operated the de-lacquering baths which were planned by the service provider.

Other main conditions of the contract are:

- delivery of the solvents and collection of the resulting output,
- list of notes,
- guarantee the quality of the solvent mixture,
- preparation and update of the safety data sheets,
- preparation of instruction documents,
- maintenance of the plant,
- use as agreed,
- complete and pure collection of the output resulting from the de-lacquering bath,
- use of the containers for collection and transport,
- avoidance of pollution of the solvents.

3. Tasks of service provider – tasks of service recipient

Due to the general prevailing conditions it was agreed that the facilities approved were to be provided and the de-lacquering was to be carried out by the staff of the service recipient. The service provider is responsible for the complete material flow necessary for the de-lacquering process.

The most important aspects are summarised in the following overview:

Service provider

- delivery of the solvents and collection of the resulting output,
- description of the measures for the bath care,
- details on notes to be taken,
- organisation of the delivery of the solvents and collection of the output,
- guarantee the quality of the solvents,
- regular analysis of the returned streams,
- preparation and update of the technical data sheets,
- preparation of instruction documents.

Service recipient

- provision of a suitable and approved storage facility,
- regular maintenance of the plants,
- reporting on the amounts of throughput and the other notes agreed on,
- compliance with the specifications for handling the solvents and output,
- use as agreed,
- complete and pure collection of the output from the de-lacquering plant,
- compliance with the specifications for the de-lacquering and bath care,

- use of the containers for the collection of the output made available by the service provider,
- avoidance of pollutions of the solvents and output,
- information in case of change of the coating-process or the powder,
- regular instruction of the employees concerned.

4. Functional unit as the basis of payment (for example: cleaned or coated surface area, hours of operation, volume of purified water, etc.) and further agreements (in particular regarding profit sharing liability issues)

At the beginning the quantity consumed was the basis of payment. It is our aim to find another basis with high significance (e.g., consumption of powder).

5. Benefits

Due to the fact that the de-lacquering facility of Mepla-Alfit was continuously undergoing further developments and that one de-lacquering bath was adapted to the de-lacquering agent of Tiefenbacher in 2004, the existing data had to be projected for one year for both baths.

The inherent data as well as the structure seem plausible because they conform to the experiences of Tiefenbacher. For clarity reasons both drawings are based on the solvent flow.

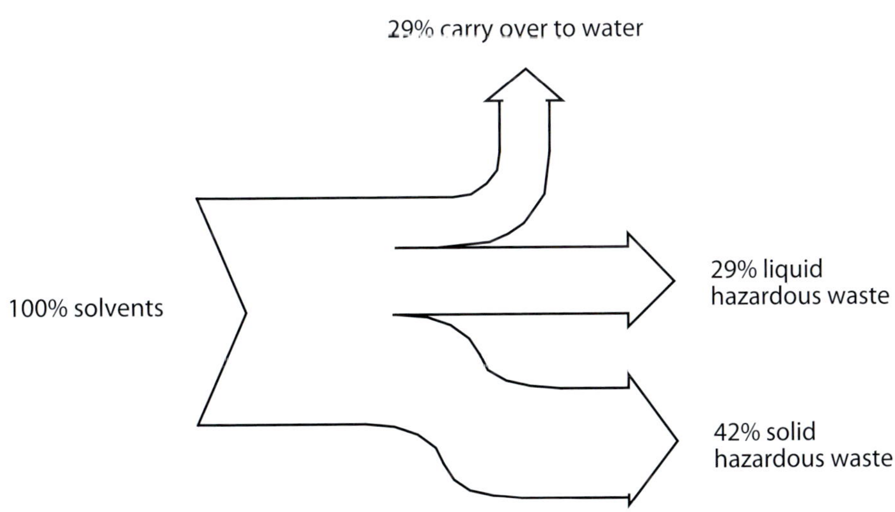

Fig. 1. Solvent flow – situation to date

The following drawing shows that about 29% of the solvent are carried into the waste water (which is fed into an internal waste water recycling). The remaining 71% can be found in the used solvent or in the filter cake and were so far disposed of as dangerous waste (Fig. 1).

Starting with the defined amount of solvent delivered, about 29% of the solvent is carried over into water, arriving at a total of about 104% hazardous waste accumulated. So far – notwithstanding the carry over of solvent – more waste was produced than solvents were introduced because also powder lacquer was found in the waste. This does not include the mass of filter cake of the internal waste water recycling.

After the implementation of Chemical Leasing the solvent balance turned out as follows (see Fig. 2).

As you see in the following figure about 50% of the solvents remain in the process. The different amounts of solvents listed as waste are due to recycling losses and the remaining solvents in the residue.

Analysis of the waste after treatment at Tiefenbacher showed that this kind of waste is not hazardous.

The mass distribution based on the solvent input is as follows (see Table 1).

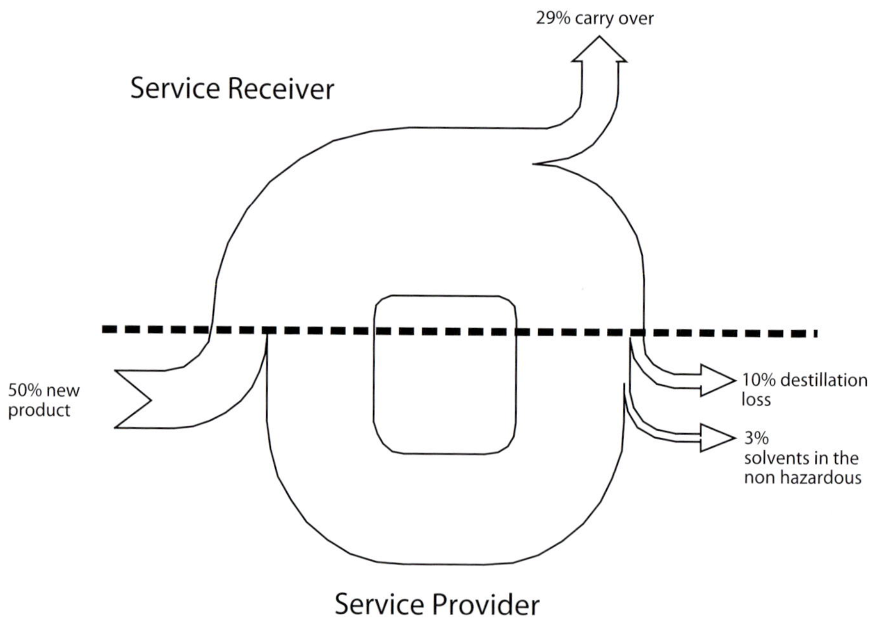

Fig. 2. Solvent flow in Chemical Leasing

Table 1. Mass distribution

	Mass distribution absolute[1)]	Mass distribution solvent
Solvent deliveries	100%	100%
Used solvent	38%	
Solvents contained therein		29%
Carry over to waste water	n.v.	29%
Solvent in filter cake		42%
Filter cake completely	66%	
Distillation residue	25%	
Solvents contained therein		3%
Loss of solvent at for instance distillation		10%

[1)] Total mass including removed powder lacquer in percent of solvents introduced.

The table shows that the necessary mass of new solvents is significantly lower in the new model of Chemical Leasing at Mepla-Alfit. The difference between the various kinds of accumulating wastes is even more prominent.

In connection with the balancing also environmental effects were recapitulated and evaluated (Table 2).

For the same service the supplier needs only appr. 80% of the total solvent consumed by the service recipient in the past. 60% of the solvent mixture are fresh solvents. The rest is treated by the supplier and reused in the process.

The conclusion is that based on the know-how of the supplier, it needs appr. 50% of fresh solvents and has to treat solvents to get the remaining 30% needed.

An additional benefit is that hazardous waste is reduced by 100%.

A list of the substantial environmental impacts shows that especially in the area of waste production significantly less impact can be achieved. The impact due to the treatment of the solvents can be neglected because the accumulated waste has also to be treated in the existing system and therefore also results in an environmental burden.

Altogether the environmental burden is reduced.

Table 2. Environmental aspects

Area	Present situation	Chemical Leasing	Evaluation
Solvent new	100% new product	ca. 50% new product necessary	Impact on environment due to production and transport is reduced
Solvent treatment	–	ca. 10% loss	Impact due to loss at distillation but omission of treatment of dangerous waste
Delivery	Shipper	Shipper	No change
Evacuation	Waste disposal	Shipper	Due to the combination of delivery and evacuation empty transports are avoided
Waste	> 100% of input	appr. 25% of input	Less environmental burden due to mass reduction and omission of dangerous waste, less quantities to be transported in the Chemical Leasing model
Treatment and disposal	Complex	Trouble-free	No strain due to waste treatment in Chemical Leasing
Research and development	Not applicable	Ongoing	De-lacquering will be further developed continuously. Increase in efficiency is aimed at and implemented
Protection of resources	–	Saving of raw material	Due to recycling only 50% new solvents are necessary

6. Experiences (Feedback asked from the participating companies by a structured questionnaire)

The companies' experiences with Chemical Leasing are definitely positive. However the implementation of a new basis of payment and agreements on price adjustments due to a fluctuating world price are still considered as problems.

Nevertheless the common goal of reducing the introduced amount of solvent was reached.

Fig. 3. Solvent treatment with a destillation plant

Questionnaire

Chemical Leasing in the Field of Paint Stripping

Fa. Tiefenbacher GmbH (Supplier)

		① LOW	②	③	④ HIGH
1	What were the key benefits expected leading to your participation in the project? **Unique offer, customer relationship, sustainability (reduction of raw material).**				
2	Have these goals been achieved?	YES **X**		NO O	
	If not, why:				
3	Has the implementation of the new business model been a challenging task in general?	① **X**	②	③	④
3	a Difficulties during/with the implementation? **Technical problems in connection with the treatment plant.**				
4	Is the new business model competitive?	①	②	③	❹
5	Are you successful with the new business model?	①	②	③	❹
6	Could you use experiences also for other products/offers?	①	②	③	❹
7	How many customers have accepted the new business model till now?	**4**			
8	Are these small (to 50 employees), medium-sized (51 to 250 employees) or large enterprises (more as 250 employees)?	**2 small/2 medium**			
9	Is the application of chemicals a core competency of these enterprises?	YES O		NO **X**	
10	Were other companies involved in the implementation of this model (equipment suppliers, disposal/recycling companies, etc.)?	YES **X**		NO O	
	If yes, please specify: **Consultant**				
11	Characteristics of chemicals involved:				
	Value	①	②	③	❹
	Recyclability	①	②	③	❹
	Hazardousness	①	②	❸	④
12	Consequences of the model:				
	Product innovation	①	②	③	❹
	Earnings	①	❷	③	④

		Cost	①	②	③	❹
		Sales volume optimized	①	❷	③	④
		Customer relationship	①	②	③	❹
		Competition	❶	②	③	④
		New customers	①	❷	③	④
		Mass of raw material/new product 50% (also in %)	①	②	❸	④

13	Is the model also suitable for very small enterprises (5 to 20 employees) or small product quantities?	YES **X**	NO O
	Reasons: **Know-how transfer, technology.**		
14	Motives for the introduction of the new business model? **Advantages in business competition.**		
15	What are the main advantages/disadvantages of this model compared to the traditional sales concept? **Customer relationship, sustainability (reduction of raw material), cost reduction.**		
16	Which charging base (unit of payment) do you use?	**kg/in the future: treated parts or consumption of powder**	
17	Has this charging base any disadvantage from your point of view?	YES O	NO **X**
	If yes, please specify: **In some cases it is difficult to explain. We had to find a simpler basis (e.g., per drum) for small enterprises.**		
18	What quantity of a new product could be saved? kg, l, %, etc.	**50%**	
19	What quantity of hazardous wastes could be reduced? kg, l, %, etc.	**100%**	
20	What was the decrease in emissions to the environment? kg, l, %, etc.	**0**	
21	Would you implement the new business model once again?	YES **X**	NO O
22	Will you use the business model also for other products?	YES **X**	NO O
23	Will this model foster the implementation of REACH?	YES **X**	NO O
	Reasons:		
24	What would you propose to change, complete or develop regarding this model?		

Chapter 5

Chemical Leasing International – Case Studies

5.1 Chemical Leasing in Egypt

5.1.1 Electrostatic Powder Coating

Ali Abo Sena, Ali Hosni, and Reinhard Joas

Case study information:

1. Information about the branch of industry and the conditions in the country

Powder coating is an effective way of protecting metal surface from deterioration. Electrostatic powder coating is a common technique used in various branches of the Egyptian industry. In this case study the supplier of electrostatic powder coating is a medium size Egyptian company with about 400 employees. It has a market share of around 60% of the Egyptian Market. The user of electrostatic powder coating is a large Egyptian company with about 2500 employees and is a subsidiary of an international group. The company is market leader in the manfacturing of electric equipment, especially high and low voltage equipment. The industry of electric technology is a very fast growing sector in Egypt. Egypt is exporting this technology to a wide range of African, Arab and Middle East countries.

The major end user sectors for electrostatic powder coating in Egypt are domestic appliances and air-conditioners manufacturers. Additionally, architectural metal finishing is also an important and growing market for the electrostatic powder coatings in Egypt. Within the electrostatic powder coating sector in Egypt there are some common problems related to manufacturing and the application of electrostatic powder. These problems are mainly the high powder losses and higher thickness of coating compared to the performance of other imported powders. In this context, there is a huge potential for improvement in the efficiency of electrostatic powder coating and for raising the competitiveness of the local industry by applying better chemical management systems.

2. Description of processes and services supplied by Chemical Leasing

Process description

The powder coating line at the user's plant consists of the following units:

- pre-treatment units (degreasing, phosphating and drying units),
- spraying booth with 8 movable spraying guns,
- control unit,
- feeding/charging unit,
- two cyclones for cleaning the booth and recycling fine powder,
- one back filter for collecting the fine powder,
- baking oven at 200 °C.

The surfaces of articles are pre-treated by degreasing in a weak alkali bath then rinsing by water shower, phosphating by dipping into phosphate solution and drying in a closed oven. The treated articles are sprayed with two sets of Tribo guns, one set is facing the exterior surface of articles, and the other set is facing the interior surface of articles. After spraying of the articles they go into the backing oven to cure the powder coat. The coated articles are collected, inspected and passed on to the assembly unit.

The production line at the electrostatic powder coating provider's plant consists of the following units:

- mixing unit,
- semi-curing unit and chipping unit,
- grinding unit,
- packing unit,
- storages for raw materials and final products.

The manufacturing process of powder coating starts with mixing the powder coating ingredients in a mixing unit (special mixers). These ingredients are mixtures of resin, pigment, filler and special additives. After the mixing process the mixed ingredients go to a semi-curing unit and are then forwarded to the chipping unit. The chips then are grinded to produce the coating powder with a specific quality.

Service provided under the Chemical Leasing business model

The supplier of electrostatic powder coating is providing the powder to the user. The powder is applied in coating the components of the power equipment (external frames of high and low voltage electric equipment). Under the

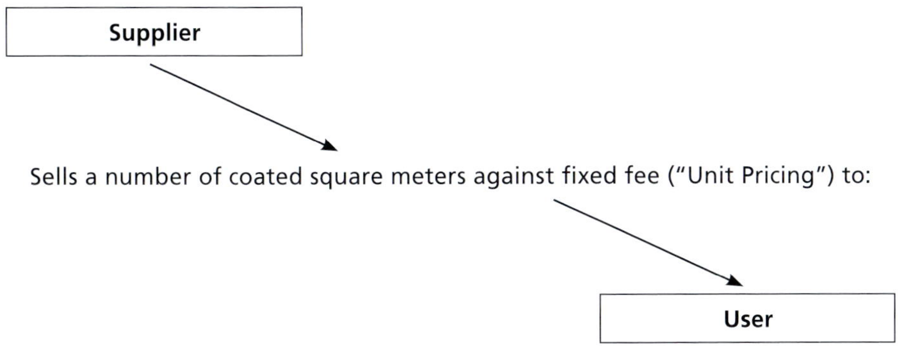

Fig. 1. Service provided according to the Chemical Leasing model

Chemical Leasing (ChL) business model, the supplier is providing the coating powder, managing and supervising the powder coating process at the suppliers plant. Furthermore, the supplier is providing a certain number of coated m^2 per month against a fixed price (see Fig. 1). The offered ChL service is including know how transfer and improvement of the quality of the final product, in addition to capacity building of the user's technical staff.

3. Tasks of service provider – tasks of service recipient and ENCPC

Tasks of Chemical Leasing service provider

- Provides coated articles of high quality in terms of square meters to the user.
- Gets back its powder losses for recycling (closing the loop).
- Trains the user's operational staff of the powder coating line.
- Organises the procedure of coating powder delivery and collection of losses in close coordination with the user.
- Provides information and data relevant to the product and its application.
- Supervises and manages the powder coating shop and establishes a reporting system to the user.

Tasks of Chemical Leasing service recipient

- Complies with the specifications for quality control.
- Provides information and data relevant to process optimisation.
- Nominates powder coating staff.

- Complies with the specifications of handling of coating powder product and outputs.
- Is committed to involve the supplier in the supervision or management of the coating powder process at its plant.
- Is responsible for the complete and pure collection of the powder losses.
- Avoids pollution of the powder coating and output.
- Complies with the operation manual standards specified by the supplier of the powder coating line.
- Informs and reports in case of failures.
- Provides a monthly report on the number of coated square meters of final product.

Tasks of UNIDO-Egypt National Cleaner Production Centre (ENCPC)

- Provides project management.
- Performs a cost and benefit analysis study on the project (feasibility in technical and economic respects).
- Support for the implementation and documentation of the project by national and international experts.
- Is responsible for process audits and recommendations for optimisation at both companies.
- Provides technical assistance to optimise and improve the efficiency of the powder coating line and the overall management.
- Is responsible for suggesting the ChL business model for the two companies and drafting the ChL agreement in close cooperation with the two companies.
- Is monitoring the implementation of the project for one year.
- Is responsible for adequate promotion, up-scaling and presentation of progress and results.

4. Functional unit as basis of payment (for example: cleaned or coated surface area, hours of operation, volume of purified water, etc.) and further agreements (in particular as to profit sharing and liability issues)

Amount (Egyptian Pounds) per coated m^2 of final product.

5. Benefits

Economical benefits

1. Saving in raw material by reduction in powder loss from 12% to 5%.
2. Cost reduction of coating process per sqm from 3.8 per sqm to 3.20 per sqm.
3. Forecast direct savings are around US$ 68 000 per year.
4. Process optimization and more efficient process (1kg powder makes more that 6 sqm rather than 5 sqm).
5. Less maintenance cost (one time instead of two times per month).
6. Less energy consumption by reduction of pressure of the powder guns.
7. Long term business relationship based on a regularly monitored contract.

Environmental benefits

1. Recycling of powder waste at Akzo Nobel (Closing the Loop).
2. Compliance with environmental regulations related to waste management and workplace environment.
3. Enhancement of supply chain management and other environmental management systems.
4. Reduction of chemicals, raw materials, energy, emissions and waste.

Organizational and quality benefits

1. Capacity building of operation staff by sharing Know-how.
2. High quality of final product ensured.

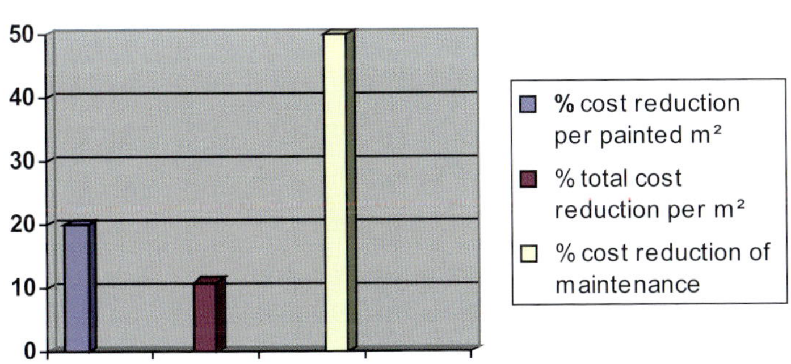

Fig. 2. Cost reduction in the Chemical Leasing business model

6. Experiences

Interviews of the participating companies by a structured questionnaire.

Please find attached the questionnaires.

7. Involved partners and country information

Supplier

- Company name: Akzo Nobel Powder Coatings S.A.E
- Field of activity: A leading company in electrostatic powder coating
- Role in the project: Supplier of electrostatic powder coating

User

- Company name: ABB ARAB
- Field of activity: Market leader in electrical technology and producer of the high and low voltage equipment
- Role in the project: User of electrostatic powder coating

Fig. 3. Chemical Leasing for powder coating at ABB ARAB

Fig. 4. Chemical Leasing for powder coating at ABB ARAB

Questionnaire

Electrostatic Powder Coating

Akzo Nobel (Supplier)

		① LOW	②	③	④ HIGH
1	What were the key benefits expected leading to your participation in the project?	high			
2	Were these goals achieved?	YES X		NO O	
	If not, why:				
3	Was the implementation of the new business model challenging in general?	①	②	③	❹
3	a Difficulties during/with the implementation? – building trust with customer – define requirements for optimisation – agreement on the basis of payment – designing of contract				
4	Is the new business model competitive	①	②	③	❹
5	Are you successful with the new business model?	①	②	③	④
6	Could you use the experiences also for other products/offers?	①	②	③	❹
7	How many customers have accepted the new business model till now?	three pipeline companies			
8	Are these small (to 50 employees), medium-sized (51 to 250 employees) or big enterprises (more as 250 employees)?	mainly medium-sized and big enterprises			
9	Is the application of chemicals a core competence of these enterprises?	YES O		NO X	
10	Were other companies involved in the implementation of this model (providers of equipment, disposal/recycling companies, etc.)?	YES X		NO O	
	If yes, please specify: Partially the supplier of the powder coating line was involved.				
11	Characteristics of involved chemicals:				
	Value	①	②	③	❹
	Recyclability	①	②	③	❹
	Hazardousness	①	❷	③	④
12	Consequences of the model:				
	Product innovation	①	②	③	❹
	Earnings	①	②	❸	④
	Cost	①	❷	③	④
	Sales volume	①	②	❸	④
	Customer relationship	①	②	③	❹

<table>
<tr><td></td><td></td><td>Competition</td><td>①</td><td>②</td><td>③</td><td>❹</td></tr>
<tr><td></td><td></td><td>New customers</td><td>①</td><td>②</td><td>❸</td><td>④</td></tr>
<tr><td></td><td></td><td>Mass of raw material/new product (also in %)</td><td>①</td><td>②</td><td>❸</td><td>④</td></tr>
<tr><td rowspan="2">13</td><td colspan="2">Is the model also suitable for very small enterprises (5 to 20 employees) or small product quantities?</td><td colspan="2">YES
X</td><td colspan="2">NO
O</td></tr>
<tr><td colspan="6">Reasons:
It depends on the value and applicability of the chemicals (great potential for very expensive chemicals used in small enterprises).</td></tr>
<tr><td>14</td><td colspan="6">Motives for the introduction of the new business model?
– higher competitiveness
– cost reduction
– strong supply chain system
– long-term business relationship</td></tr>
<tr><td>15</td><td colspan="6">What are the main advantages/disadvantages of this model as compared to the traditional sales concept?
Advantages:
– sharing benefits
– higher competitiveness
– product Innovation
– quality and environment benefits
Disadvantages:
– high trust in the customer is required
– third party for monitoring implementation is very essential</td></tr>
<tr><td>16</td><td colspan="2">Which charging base (unit of payment) do you use?</td><td colspan="4">per square meter of coated article</td></tr>
<tr><td rowspan="2">17</td><td colspan="2">Has this charging base any disadvantage from your point of view?</td><td colspan="2">YES
O</td><td colspan="2">NO
X</td></tr>
<tr><td colspan="6">If yes, please specify:</td></tr>
<tr><td>18</td><td colspan="2">What quantity of a new product could be saved?
kg, l, %, etc.</td><td colspan="4">20% of new product</td></tr>
<tr><td>19</td><td colspan="2">What quantity of hazardous wastes could be reduced?
kg, l, %, etc.</td><td colspan="4">4% of waste will be recycled</td></tr>
<tr><td>20</td><td colspan="2">What was the decrease of emissions to the environment?
kg, l, %, etc.</td><td colspan="4">reduction in waste from 12% to 4% and recycling of this waste at the supplier plant</td></tr>
<tr><td>21</td><td colspan="2">Would you implement the new business model once again?</td><td colspan="2">YES
X</td><td colspan="2">NO
O</td></tr>
<tr><td>22</td><td colspan="2">Will you use the business model also for other products?

We might be user of the ChL services in cooperating with our suppliers.</td><td colspan="2">YES
X</td><td colspan="2">NO
O</td></tr>
<tr><td rowspan="2">23</td><td colspan="2">Will this model foster the implementation of REACH?</td><td colspan="2">YES
X</td><td colspan="2">NO
O</td></tr>
<tr><td colspan="6">Reasons:
Might be right if we are exporting to the EU market.</td></tr>
<tr><td>24</td><td colspan="6">What would you change, complete or develop in this model?
I would work and cooperate on the development of a standard certification scheme for ChL. Further promotion of this concept through a global partnership between Akzo Nobel and UNIDO.</td></tr>
</table>

Questionnaire

Electrostatic Powder Coating

ABB Arab (User)

			① LOW	②	③	④ HIGH
1	What were the key benefits expected leading to your participation in the project?		④			
2	Were these goals achieved?		YES **X**		NO O	
	If not, why:					
3	Was the implementation of the new business model challenging in general?		①	②	③	❹
3	a	Difficulties during/with the implementation? • **define and approve the requirements for optimisation between the parties** • **design the Chemical Leasing contract and define the responsibilities** • **lack of a reference for a success story** • **high trust between involved parties was required**				
4	Does the new business model provide cost reductions?		①	②	③	❹
5	Are you successful with the new business model?		①	②	③	❹
6	Were other companies involved in the implementation of this model (providers of equipment, disposal/recycling companies, etc.)?		YES **X**		NO O	
	If yes, please specify: **The supplier of the powder coating line was involved at the beginning to elaborate the standard measures of operation.**					
7	Characteristics of involved chemicals:					
		Value	①	②	❸	④
		Recyclability	①	②	③	❹
		Hazardousness	①	❷	③	④
8	Consequences of the model:					
		Product innovation	①	②	③	❹
		Earnings	①	②	③	❹
		Cost	①	❷	③	④
		Supplier relationship	①	②	③	❹

	Competition	①	②	③	❹
	Mass of raw material/new product (also in %)	①	②	❸	④
9	Is the model also suitable for very small enterprises (5 to 20 employees) or small product quantities?	YES X		NO O	
	Reasons: Might be, taking into consideration the cost of these chemicals.				
10	Motives for the introduction of the new business model? • cost reduction • more efficient process • high quality product • well skilled staff • less liability • environmental benefits				
11	What are the main advantages/disadvantages of this model as compared to the traditional sales concept? Advantages: • provides a significant cost reduction • leads to more efficient process • enhancement of the supply chain management of the company by building a strong relationship with suppliers • high environmental benefits				
12	Which charging base (unit of payment) do you use?	per square meter of coated articles (final product)			
13	Has this charging base any disadvantage from your point of view?	YES X		NO O	
	If yes, please specify:				
14	What quantity of a new product could be saved? kg, l, %, etc.	20%			
15	What quantity of hazardous wastes could be reduced? kg, l, %, etc.	the fine powder waste will be recycled at AKZO NOBEL.			
16	What was the decrease of emissions to the environment? kg, l, %, etc.	100%			
17	Would you implement the new business model once again?	YES X		NO O	
18	Will you use the business model also for other products?	YES X		NO O	
19	Will this model foster the implementation of REACH?	YES X		NO O	
	Reasons: Because we are exporting to the EU countries.				
20	What would you change, complete or develop at this model? Put down a standard procedure for Chemical Leasing implementation and promotion of success stories.				

5.1.2 Cleaning Equipment with Hydrocarbon Solvent

Ali Abo Sena, Ali Hosni, and Reinhard Joas

Case study information:

1. Information about the branch of industry and the conditions in the country

The process of cleaning of painting equipment in different industrial sectors is very important to ensure that the painting work will be done efficiently and with high quality. Hydrocarbon solvents are used to dissolve substances which cannot be dissolved in water. Reducing the consumption of solvent and its wastes is still one of the most challenging elements of pollution prevention worldwide. Solvent use is a major contributor to workplace environment pollution (including ambient air), water pollution and is considered a source of hazardous waste. Proper management of solvent is essential as solvents can cause fires, explosions, contamination of water, soil and workplace environment, and can be harmful to human health.

Solvents for cleaning of painting equipment are widely used in the Egyptian industry, especially the automotive, pharmaceutical and printing industries. Unfortunately, there is a lack of proper and environmentally sound management of solvent wastes which are considered as hazardous wastes.

In the present case the supplier of the hydrocarbon solvent is a small-sized private Egyptian company, highly specialised in chemicals used in metal finishing. Among others, this company produces chemicals for phosphating, electroplating, cleaning with hydrocarbon solvents, pickling and degreasing processes. 70% of its raw materials are locally produced and 30% are imported.

The user in this present case is a large Egyptian company and subsidiary of an international group. Egypt has around 15 working national and international automotive companies and most of them are located in 6th of October city. However, the user's facility is the largest vehicle assembly plant in Egypt. As most of the automotive companies, they are using the hydrocarbon solvent in the cleaning of equipment especially in its painting shops.

The dramatic increase in the petroleum prices has lead to an increase of prices of the raw materials for production of hydrocarbon solvent. Additionally, the environmental problems related to air emission and liquid waste from using these hydrocarbon solvent were the driving forces for new and innovative models for managing such kind of chemicals.

As the problem of management of solvent waste is a common problem within the automotive sector in Egypt, this case might be very interesting to other automotive companies to join a regional project on solvent management.

2. Description of processes and services supplied by Chemical Leasing

Process description

The supplier of hydrocarbon solvent prepares the solvent by blending different solvents and organic compounds. The company does not disclose the exact composition of the solvent, only stating that it is a mixture of toluene, butanol and ethanol with a 40–50% toluene content. The different constituents are bought at the local market. The prepared solvent contains traces of water and heavy residues.

The cleaning solvent is tested for its capacity to dissolve paints. The recycled solvent will require some analyses to allow an estimation of the required amounts of different constituents that need to be added in order to reformulate the fresh solvent.

The car painting operation at the automotive company is the unit that generates VOC emissions and solvent waste. There are two paint booths each followed by a curing oven. The cars are mechanically introduced into the booths where they are spray painted. The spray booths are equipped with an underground water curtain that collects excess paint. Furthermore, the ventilation is such that air with its VOC content is forced downward through the water curtain which acts as a scrubber. According to the user's personnel the VOC discharged through the stacks of the spraying booths is within permissible limits. The solvent is used in three main operations:

- **Cleaning of guns and connecting piping**: The quality of a car painting being a major factor in the customer's decision to buy which car, the company works hard to keep the quality of the final finish high. Besides applying the paint correctly and well, the careful cleaning of the paint spraying equipment and changing the associated piping each time the color of paint is applied also are important factors. This is necessary to ensure that a given colour is uniform and consistent in each vehicle of that colour. The solvent is placed in a barrel next to the paint container which is connected to the spraying equipment by tubing. When purging of the line is required, the tube is unplugged from the paint barrel and plugged to the solvent barrel. The purged paint and cleaning solvent are collected in a separate container placed at the spray booth. The containers of waste solvent are stored in a specified area for hazardous waste. They are presently being sold to a sub-contractor in closed barrels for LE 1/bbl. The amount of VOCs escaping from this operation is negligible as it is performed in a closed cycle.

- **Sealing**: A major part of the cleaning solvent is used with a cloth to clean the cars after paint spraying. The solvent is completely evaporated and cannot be recovered since sealing takes place outside the paint booth.
- **Cleaning of tanks**: Tanks containing the paints are cleaned only when a batch of colour has been discontinued or found defective. This requires large amounts of solvent but does not happen often (once every 6–8 months).

Services provided by the Chemical Leasing business model

The service provider is supplying the hydrocarbon solvent and supervises and gives recommendations for using the solvent in the process of cleaning equipment and takes back the solvent waste for recycling.

3. Tasks of service provider – tasks of service recipient and ENCPC

Tasks of Chemical Leasing service provider

- Providing technical data on the solvent specifications and its application guideline.
- Providing the required Material Safety Data Sheet (MSDS) for the hydrocarbon solvent used in the cleaning process.
- Know-how transfer to the user, especially in handling and application of hydrocarbon solvent in the cleaning processes of equipment.
- Giving recommendations for improvement of the processes of cleaning the equipment at the user's plant.
- Collecting the solvent waste from the cleaning processes and delivering for recycling at the service provider's facility and laboratory.
- Supplying 100% of the solvent required during the term of the agreement.

Tasks of Chemical Leasing service recipient

- As user of the hydrocarbon solvent committed to comply with the technical specifications for the cleaning process of equipment.

- Committed to provide to the service provider full information and all data relevant to the process of cleaning with hydrocarbon solvent.
- Nomination of the cleaning staff to be supervised and followed up regularly by the service provider.
- Providing a storage place for hydrocarbon solvent.
- Complying with the specifications of handling and application of hydrocarbon solvent which are provided.
- Committed to involve the service provider in following up the equipment of cleaning process during the term of the agreement.
- Committed to cooperate in developing a regular process of reporting to the supplier on the number and type of cleaning processes on a monthly basis.
- Ensuring that the cleaning process is staffed by suitably skilled personnel.

Tasks of UNIDO-Egypt National Cleaner Production Center (ENCPC)

- Provides the project management.
- Performs a cost and benefit analysis study of the project (feasibility in technical and economic respect).
- Support for implementation and documentation of the project by national and international experts.
- Is responsible for process audits and recommendations for optimisation at both companies.
- Provides technical assistance to optimise and improve the efficiency of powder coating line and overall management.
- Is responsible for suggesting the ChL business model between the two companies and drafting the ChL agreement in close cooperation with the two companies.
- Is monitoring the implementation of the project for one year.
- Is responsible for adequate promotion, up-scaling and presentation of progress and results.

4. Functional unit as basis of payment (for example: cleaned or coated surface area, hours of operation, volume of purified water, etc.) and further agreements (in particular as to profit sharing and liability issues)

Amount (Egyptian Pound) per produced vehicle.

5. Benefits

Economical benefits

1. Cost reduction by 15% (saving of raw material with recycling).
2. Reduction of solvent consumption from 1.5 L per vehicle to 1 L per vehicle.
3. Share liability and benefits.
4. Ensure long term business relationship based on long contract.
5. Saving the cost of getting rid from the solvent waste.

Environmental benefits

1. Recycling of solvent waste and stop dumping (Closing the Loop).
2. Better hazardous waste management in accordance to environmental regulations and international environmental corporate policy.

Organizational and management benefits

1. More efficient cleaning process with hydrocarbon solvent by applying a procedure for cleaning by batch cleaning.
2. Enhancement of supply chain management and other environmental management systems.
3. Stop of using the hydrocarbon solvent in other purposes rather that cleaning of equipment (e.g., stop washing worker hands, cloths etc.).
4. Capacity building and awareness of operation staff.

6. Experiences

Interviews of the participating companies with a structured questionnaire.

Please find attached the questionnaires.

7. Involved partners and country information

Supplier

- Company name: Dr. Badawi Chemical Work
- Field of activity: Chemicals used in metal finishing
- Role in the project: Supplier of hydrocarbon solvent

User

- Company name: GM Egypt
- Field of activity: Automotive company (engineering)
- Role in the project: User of hydrocarbon solvent

Fig. 1. Solvents preparation at Dr. Badawi

Fig. 2. Solvents used for Chemical Leasing at General Motors

Questionnaire

Cleaning Equipment with Hydrocarbon Solvent

Dr Badawi (Supplier)

			① LOW	②	③	④ HIGH
1		What were the key benefits expected leading to your participation in the project?				
2		Were these goals achieved? The project will be monitored for the first one year.	YES X		NO O	
		If not, why:				
3		Was the implementation of the new business model challenging in general?	①	②	③	❹
3	a	Difficulties during/with the implementation? • the need to purchase a recycling unit • investigation of the possibility to involve other six automotive companies to maximize the benefits • define the most adequate basis of payment for the Chemical Leasing business model				
4		Is the new business model competitive?	①	②	③	❹
5		Are you successful with the new business model?	①	②	❸	④
6		Could you use the experiences also for other products/offers?	①	②	③	④
7		How many customers have accepted the new business model till now?	There is a possibility to involve another six automotive companies in the new model (Chemical Leasing company). Additionally, we have two clients in the electroplating sector and aluminium sector who are ready to cooperate in ChL projects.			
8		Are these small (to 50 employees), medium-sized (51 to 250 employees) or big enterprises (more as 250 employees)?	mainly medium-sized and big companies			
9		Is the application of chemicals a core competence of these enterprises?	YES O		NO X	
10		Were other companies involved in the implementation of this model (providers of equipment, disposal/recycling companies, etc.)?	YES O		NO X	
		If yes, please specify:				
11		Characteristics of involved chemicals:				
		Value	①	②	❸	④
		Recyclability	①	②	③	❹
		Hazardousness	①	②	③	❹
12		Consequences of the model:				
		Product innovation	①	②	③	❹

	Earnings	①	②	③	❹
	Cost	❶	②	③	❹
	Sales volume	①	②	❸	④
	Customer relationship	①	②	③	❹
	Competition	①	②	③	❹
	New customers	①	②	③	❹
	Mass of raw material/new product (also in %)	①	②	③	❹
13	Is the model also suitable for very small enterprises (5 to 20 employees) or small product quantities?	YES X		NO O	
	Reasons: This might be achieved if this small enterprise is co-operating with other big companies in a multi-company Chemical Leasing project (e.g. ChL for the cleaning of equipment with solvent in automotive companies). For instance the cleaning shops of machine parts, being very small shops, might be involved in such a model.				
14	Motives for the introduction of the new business model? • long business relationship with GM Egypt • competitiveness • profitability • environmental benefits				
15	What are the main advantages/disadvantages of this model as compared to the traditional sales concept? Advantages: • high potential for cost saving in colouring by more efficient process • environmental benefits Disadvantages: • needs high trust between partners • more responsibility for the supplier • an independent monitoring expertise body is required				
16	Which charging base (unit of payment) do you use?	per vehicle produced			
17	Has this charging base any disadvantage from your point of view?	YES O		NO X	
	If yes, please specify:				
18	What quantity of a new product could be saved? kg, l, %, etc.	10–15%			
19	What quantity of hazardous wastes could be reduced? kg, l, %, etc.	80–100% of hazardous waste			
20	What was the decrease of emissions to the environment? kg, l, %, etc.				
21	Would you implement the new business model once again?	YES X		NO O	
22	Will you use the business model also for other products?	YES X		NO O	
23	Will this model foster the implementation of REACH?	YES X		NO O	
	Reasons: Yes, in case we are exporting or our customers are exporting to the EU countries.				
24	What would you change, complete or develop in this model? I would work in a Chemical Leasing company in cleaning with hydrocarbon solvent with involvement of other companies from different sectors. At the same time I would like to work on the development a standard scheme for certification of Chemical Leasing.				

Questionnaire

Cleaning Equipment with Hydrocarbon Solvent

GM Egypt (User)

		① LOW	②	③	④ HIGH
1	What were the key benefits expected leading to your participation in the project?	high			
2	Were these goals achieved?	YES X		NO O	
	If not, why:				
3	Was the implementation of the new business model challenging in general?	①	②	③	❹
3 a	Difficulties during/with the implementation? • define the basis of payment • long procedure to design the contract • lack of case studies				
4	Does the new business model provide cost reductions?	①	②	③	❹
5	Are you successful with the new business model? The ChL contract is just in place and there is a very high potential for success.	①	②	③	❹
6	Were other companies involved in the implementation of this model (providers of equipment, disposal/recycling companies, etc.)?	YES O		NO X	
	If yes, please specify:				
7	Characteristics of involved chemicals:				
	Value	①	❷	③	④
	Recyclability	①	②	③	❹
	Hazardousness	①	②	③	❹
8	Consequences of the model:				
	Product innovation	①	②	③	❹
	Earnings	①	②	③	④
	Cost	❶	②	③	④
	Supplier relationship	①	②	③	❹
	Competition	①	②	③	❹
	Mass of raw material/new product (15–20%)	①	②	③	❹
9	Is the model also suitable for very small enterprises (5 to 20 employees) or small product quantities?	YES X		NO O	
	Reasons: According to the volume of solvent used and waste produced.				

<table>
<tr><td>10</td><td colspan="3">Motives for the introduction of the new business model?
• cost reduction
• efficient cleaning process
• environmental benefits and meeting local and international environmental corporate policies
• decrease liability</td></tr>
<tr><td>11</td><td colspan="3">What are the main advantages/disadvantages of this model as compared to the traditional sales concept?
Advantages:
• cost reduction
• safe management of hazardous wastes
• reduction of the consumption of solvent and higher efficient process (more control over the process)
• combining the aspects of better process performance and low cost with environmental aspects</td></tr>
<tr><td>12</td><td>Which charging base (unit of payment) do you use?</td><td colspan="2">per vehicle produced</td></tr>
<tr><td rowspan="2">13</td><td>Has this charging base any disadvantage from your point of view?</td><td>YES
O</td><td>NO
X</td></tr>
<tr><td colspan="3">If yes, please specify:</td></tr>
<tr><td>14</td><td>What quantity of a new product could be saved?
kg, l, %, etc.</td><td colspan="2">15–20%</td></tr>
<tr><td>15</td><td>What quantity of hazardous wastes could be reduced?
kg, l, %, etc.</td><td colspan="2">100%</td></tr>
<tr><td>16</td><td>What was the decrease of emissions to the environment?
kg, l, %, etc.</td><td colspan="2">around of 15%</td></tr>
<tr><td>17</td><td>Would you implement the new business model once again?</td><td>YES
X</td><td>NO
O</td></tr>
<tr><td>18</td><td>Will you use the business model also for other products?</td><td>YES
X</td><td>NO
O</td></tr>
<tr><td rowspan="2">19</td><td>Will this model foster the implementation of REACH?</td><td>YES
X</td><td>NO
O</td></tr>
<tr><td colspan="3">Reasons:
It might influence the implementation of REACH where we are part of a global company and much concerned by these issues.</td></tr>
<tr><td>20</td><td colspan="3">What would you change, complete or develop in this model?
I would suggest to add more companies participating in this model, especially for the implementation of Chemical Leasing in the cleaning with hydrocarbon solvent, where it is essential to have a standard Chemical Leasing procedure for implementation when we start new cases.</td></tr>
</table>

5.1.3 Hot Dip Galvanisation

Ali Abo Sena, Ali Hosni, and Reinhard Joas

Case study information:

1. Information about the branch of industry and the conditions in the country

Currently, there are about seven big Egyptian companies specialised in zinc galvanisation. Two galvanisation technologies are generally used: Hot Dip Galvanisation and Continuous Sheet Galvanisation. In Hot Dip Galvanisation, the structural tower parts, fasteners, street electricity transmission towers, etc. are galvanised to give a durable, maintenance free product while in Continuous Sheet Galvanisation only metal sheets are galvanised. Only one company in Egypt is specialised in Sheet Galvanisation, the others are specialised in Hot Dip Galvanisation. Generally, the galvanisation is done based on specific national requirements (e.g., thickness of zinc layer, silica content, etc.)

2. Description of processes and services supplied by Chemical Leasing

Process description

The following flowchart summarises the Hot Dip Galvanisation process (see Fig. 1).

The flowchart of Fig. 2 describes the processes linked to the galvanisation process at the company supplying the flux.

Service provided according to the Chemical Leasing business model

In Chemical Leasing, the supplier will provide the service of the fluxing process (using ammonium chloride and zinc chloride) to the user (see Fig. 3).

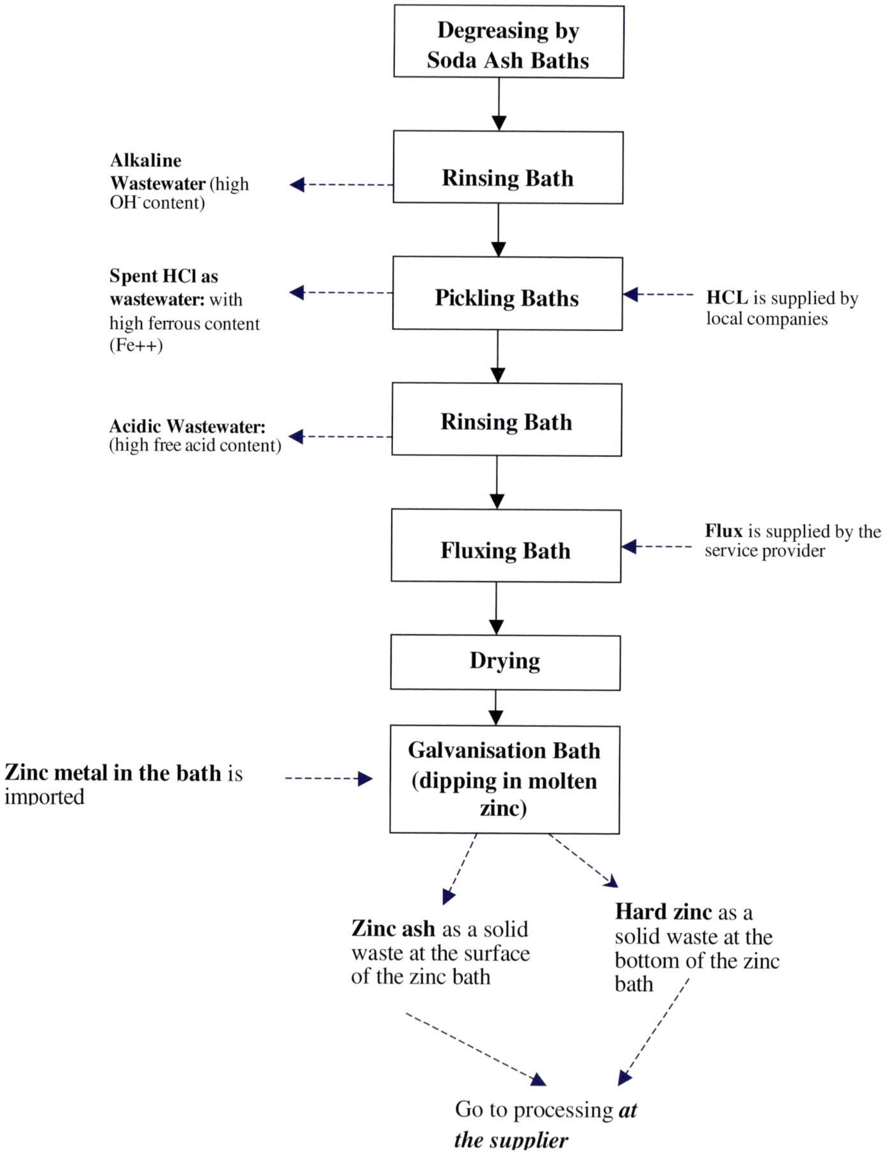

Fig. 1. Hot Dip Galvanisation process

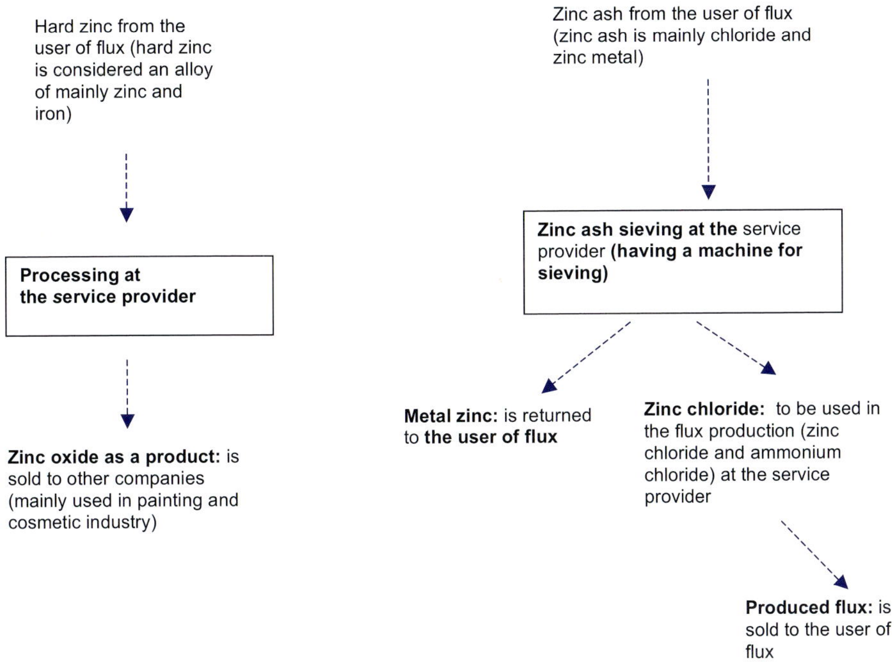

Fig. 2. Use of hard zinc and zinc ash

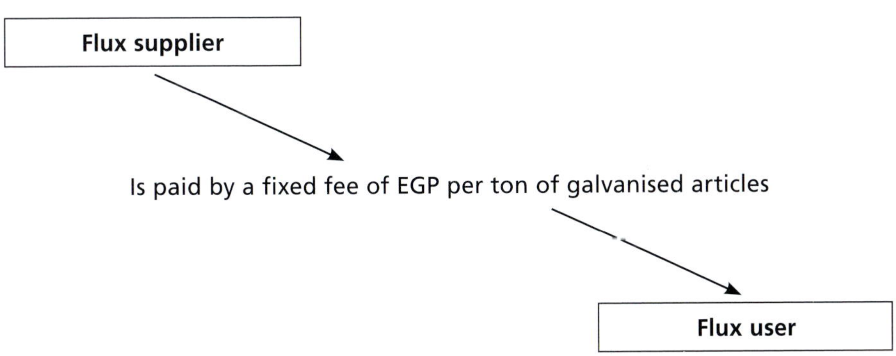

Fig. 3. Service provided according to the Chemical Leasing business model

3. Tasks of service provider – tasks of service recipient and the ENCPC

Tasks of Chemical Leasing service provider

- Is committed to share know-how and experience related to flux application at the user's plant.
- Supplies sufficient quantities of flux, in high quality (as agreed).
- Is commited to take care of the recovery of used flux, while the recovery process will take place at the user's facility.
- Supervises and manages the fluxing process including creation of a reporting system and preparation of instruction documents for the application of the flux at the user's plant.
- Organises the delivery of the flux and collection of used flux in close coordination with the user.
- Provides information and data relevant to the flux and its application (including technical data sheet).
- Assigns and nominates a person to be responsible for supervising the fluxing process at the user's plant.
- Helps in preparation and adjustment of the flux in form of a liquid solution (which requires the mixture and liquefaction of ammonium chloride and zinc chloride in a specific ratio).
- Will supply 100% of the quantities of the required flux to the user during the term of the agreement.

Tasks of Chemical Leasing service recipient

- Complies with the technical specifications in the operation manual of pickling and fluxing line.
- Provides information and data relevant to the process monitoring and optimisation of the fluxing bath.
- Nominates the fluxing process staff to be supervised and followed up by the service provider.
- Involves the supplier in supervising and following up the fluxing process during the term of the agreement.
- Receives a number of fluxed articles at the payment basis of tons of galvanised articles per time unit.
- Cooperates in developing a regular reporting process to the service provider on the number and tons of galvanised articles per month.
- Avoids pollution and contamination of the fluxed articles and other output.

- Provides information and reporting in case of failures.
- Makes containers for flux available to serve as stand-by containers in the case of recovery of spent flux.

Tasks of UNIDO-Egypt National Cleaner Production Centre (ENCPC)

- Is responsible for project management.
- Performs a cost and benefit analysis study for the project (feasibility in technical and economic respect).
- Support for implementation and documentation of the project by national and international experts.
- Is responsible for process audits and recommendations for optimisation at both companies.
- Provides technical assistance to optimise and improve the efficiency of powder coating line and overall management.
- Is responsible for suggesting the ChL business model to both companies and drafting the ChL agreement in close cooperation with the two companies.
- Is monitoring the implementation of the project for one year.
- Is responsible for adequate promotion, up-scaling and presentation of progress and results.

4. Functional unit as basis of payment (for example: cleaned or coated surface area, hours of operation, volume of purified water, etc.) and further agreements (in particular as to profit sharing and liability issues)

Amount (Egyptian Pound) per galvanised ton of final product.

5. Forecast benefits

Economic benefits

1. Expecting direct annual benefit of US $ 200,000.
2. Significant reduction in hard zinc and zinc ash production (cost reduction due to reduction in the consumption of zinc).

Environmental benefits

1. Recovery of flux waste and stop discharging to sewer system.

2. Closing the loop of solid waste (Zinc ash recovery and hard Zinc recycling).
3. Better workplace conditions.

Quality and management benefits

1. Better performance and higher efficiency of galvanization process.
2. Improve the user's productivity by concentrating labour force in the core business of the company.
3. Enhancement of supply chain management and other environmental management systems.
4. Skilled and well trained operational staff at the user's plant in fluxing process.

6. Experiences

Interviews of the participating companies by a structured questionnaire.

7. Involved partners and country information:

Supplier

- Company name: Zinc Misr Company
- Field of activity: Small private company, focusing its business on management and recovery from different industries
- Role in the project: Flux supplier (ammonium chloride and zinc chloride)

User

- Company name: El Sewedy Company
- Field of activity: Egyptian market leader in Hot Dip Galvanisation industry (zinc galvanisation), specialised in galvanising towers for power transmission, lighting, communication, etc.
- Role in the project: Flux user in the fluxing process

Fig. 4. Chemical Leasing at El Sewedy Company

Fig. 5. Chemical Leasing at El Sewedy Company

5.2 Chemical Leasing in Mexico

5.2.1 Sugar Mills

Elisa Arreola Valerio, Jorge Perez, Ignacio Sanchez, and Reinhard Joas

Case study information:

1. Information about the branch of industry and the conditions in the country

The sugar cane industry in Mexico represents an important source of generation of income and direct and indirect jobs for an important sector of the population that is located in 15 states of the Mexican Republic.

The industry as a whole accounts for more than 300,000 jobs, including cane cutters, seasonal field workers, and factory workers. Consequently, over 2.2 million people depend on the Mexican sugar industry for a living. There are 158,000 cane growers averaging nearly 4 hectares per grower delivering about 300 tons of cane.

With 59 sugar mills and nearly 160 thousand sugar cane producers, this industry registers an annual production average of five million tons of sugar, which is equivalent to two thousand five hundred million dollars, reaching a national impact that is translated in 0.4 percent of the gross internal product of the country. Due to the high proportion of the gross internal product combined with the big consumption of sugar, directly or as a raw material, the sugar production and the price level are strategic to favour the well-being of big sectors of the population and the competitiveness of diverse industries of the country.

Operational and financial problems of the sugar mills

- Low volume of cane available to mill, which implies a smaller use of its installed capacity of production.
- Low saccharine content in the cane to obtain an efficient production.
- Extraordinary losses in the saccharine extraction during the production process.
- Excessive costs of manual labour in production and maintenance of sugar mills.
- High financial costs derived from refinancing of losses of previous cycles.

- The great deterioration of the financial conditions of many sugar mills makes the necessary expansion economically nonviable.
- The production of the industry is insufficient to cover the internal demand, which implies to satisfy it with sugar imports.
- The sale prices of sugar are insufficient to cover the costs of the product.

For sugar production various chemicals are needed and this sector is of major importance for Mexico and other Latin American countries.

The sugar mill in this case study is the most important sugar producer in Mexico and produces milled sugar cane of around 3 million tons per zafra (6 months of work period). This sugar mill is part of FEESA (Association of sugar producers) together with other 20 sugar mills in Mexico.

2. Description of processes and services supplied by Chemical Leasing

The use of lubricants in sugar mills is an example for existing Chemical Leasing potentials. Here negotiations between 8 different companies took place. All of them offered work using the new project Chemical Leasing for the zafra 2007, applying greases in the mill area (bearings and disk dent mills, open transmissions).

The different lubricant suppliers made a technical proposal of their services (technical approach, monitoring, quality of the products, take back lubricants exhaust from the sugar mill, reports on lubricant consumption, environmental benefits like: water consumption reduction, energy savings, increase of equipment life-cycles, statistics and controlling) as well as a time schedule for the work plan. Each of the suppliers accepted agreements concluded in a collaboration contract between the Mexican Cleaner Production Centre, the sugar mill and the lubricants supplier.

It is important to mention that FEESA's sugar mills are a public group. Therefore, when sugar mills need suppliers, they start a selection process to obtain the best option (in Mexico this process is named "Licitación"). The process consists of an examination of the legal, technical and cost related conditions offered by each supplier.

The work of the Mexican Cleaner Production Centre was to support the technical analysis together with the sugar mill to identify the best services, product quality and environmental benefits offered by different suppliers. After the evaluation the suppliers for the zafra were selected, and a contract was signed to implement Chemical Leasing. For the zafra 2007 two supplying companies were selected.

One offers oils and other greases to be applied in different process areas in the sugar mill industry. There are around 20 different products, and they are applied in different points like pumps, sugar cane conveyors, bearing turbogenerators, speed reducers and others. The aim is to identify opportunities to apply Chemical Leasing in other points of the process, and identify the payment index (for example $ per work hours).

3. Tasks of service provider – tasks of service recipient

Tasks of service provider

- Supply of greases with biodegradability certificates and nontoxic degree by NSF H1 (National Sanitation Foundation) and FDA 21 CFR (Food and Drug Administration).
- Inspection of work conditions in the production plants, offering to have a technician in the plant every day.
- Preparation of weekly product and service reports.
- Used oil analyses every 2 months in zafra.
- Lubrication training (2 in zafra).

Tasks of service recipient

- In order to facilitate the implementation and evaluation of the new business scheme, the infrastructure and information required by the service provider will be defined.
- In the areas officially agreed, exclusively the chemical products the service provider offers to the Sugar Mill are used as raw materials for the process that is tried to be established.
- To provide all the personnel necessary to carry out the modification of the technology on site.
- To consider the recommendations given by the supplier on the correct use and application of greases, in order to always operate in optimal conditions and considering the variables of the process.
- To have a comprehensive knowledge of the state of the process ensuring immediate solution of any technical problem of operation that might exist or contacting the service provider immediately.

Both user and supplier

- To elaborate a work plan.
- To collaborate with the group of the Mexican Cleaner Production Centre to comply with the business model.

- To implement adaptations that arise as a result of the evaluation, and continue promoting and fomenting the principles and practices of Chemical Leasing, according to its possibilities of continuous improvement.

4. Functional unit as basis of payment (for example: cleaned or coated surface area, hours of operation, volume of purified water, etc.) and further agreements (in particular as to profit sharing and liability issues)

$ (pesos) per ton sugar cane milled (grease applied in mill area).

5. Benefits

- Increase of process efficiency;
- Chemical products more environmentally friendly;
- Water and energy consumption decrease;
- Waste reduction;
- Increase in production time, because the new product can help to reduce time losses due to equipment repair;
- Increase in equipment life-cycles by reducing machinery wear (frictions reduction);
- Technical support and specialised services;
- Pay by productivity.

6. Experiences

Interviews of the participating companies by a structured questionnaire.

Please find attached the questionnaires.

7. Involved partners

- Company name: Fideicomiso Ingenio San Cristóbal 80333
- Field of activity: sugar mill
- Role in the project: service receiver

- Company name: Chemical Mac Oil, S.A. de C.V. (Schutz Oil)
- Field of activity: Lubricants supplier
- Role in the project: Service provider

- Company name: Suministro de Materiales Industriales, S.A. de C.V. (SUMAT)
- Field of activity: Lubricants supplier
- Role in the project: Service provider

Fig. 1. Gear where lubricants are used

Fig. 2. Sugar mill equipment

Questionnaire

Sugar Mills

Chemical Mac Oil (Supplier)

			① LOW	②	③	X HIGH
1	What were the key benefits expected leading to your participation in the project? To offer our know how on process optimisation related to lubrification in a pilot project to a sugar mill and extend business to various sugar mills in the region.					
2	Were these goals achieved?		YES		NO X	
	If not, why: The goals could be achieved in part, but due to technical problems an extension of the business model to other sugar mills could not be achieved.					
3	Was the implementation of the new business model challenging in general?		①	②	③	❹
3	a	Difficulties during/with the implementation? Problems with technical defects of the equipment; various parameters of the incoming sugar cane and the conditions at the mills caused difficulties with the lubrification.				
4	Is the new business model competitive		①	②	❸	④
5	Are you successful with the new business model?		❶	②	③	④
6	Could you use the experiences also for other products/offers?		①	❷	③	④
7	How many customers have accepted the new business model till now?		1			
8	Are these small (to 50 employees), medium-sized (51 to 250 employees) or big enterprises (more as 250 employees)?		high			
9	Is the application of chemicals a core competence of these enterprises?		YES		NO X	
10	Were other companies involved in the implementation of this model (providers of equipment, disposal/recycling companies, etc.)?		YES O		NO X	
	If yes, please specify:					
11	Characteristics of involved chemicals:					
		Value	①	❷	③	④
		Recyclability	①	②	❸	④
		Hazardousness	①	❷	③	④
12	Consequences of the model:					

		Product innovation	❶	②	③	④
		Earnings	①	❷	③	④
		Cost	①	②	③	❹
		Sales volume	①	❷	③	④
		Customer relationship	①	②	③	❹
		Competition	①	②	③	❹
		New customers	①	❷	③	④
		Mass of raw material/new product (also in %)	①	❷	③	④

13	Is the model also suitable for very small enterprises (5 to 20 employees) or small product quantities?	YES X	NO
	Reasons:		
14	Motives for the introduction of the new business model? Condition of the tendering procedure		
15	What are the main advantages/disadvantages of this model over the traditional sales concept? Advantages: It is possible to reduce lubrification due to specific know how. Disadvantages: Many parameters to be considered that are not in the influence of the provider.		
16	Which charging base (unit of payment) do you use?	Pesos/t milled sugar cane	
17	Has this charging base any disadvantage from your point of view?	YES	NO X
	If yes, please specify:		
18	What quantity of a new product could be saved? kg, l, %, etc.	50%	
19	What quantity of hazardous wastes could be reduced? kg, l, %, etc.	25%	
20	What was the decrease of emissions to the environment? kg, l, %, etc.	25%	
21	Would you implement the new business model once again?	YES X	NO O
22	Will you use the business model also for other products?	YES X	NO
23	Will this model foster the implementation of REACH?	YES O	NO X
	Reasons: REACH has no consequences for Mexican sugar mills.		
24	What would you change, complete or develop at this model? Additional parameters have to be considered for the next production period.		

Questionnaire

Sugar Mills

Fideicomiso Ingenio San Cristóbal (User)

		① LOW	②	③	X HIGH
1	What were the key benefits expected leading to your participation in the project? • reduced consumption of lubricants • cost reduction • increase of the life-cycles of the equipment/reduced damages • obtain HACCP • reduced lubricants emission to the environment				
2	Were these goals achieved?	YES X		NO X	
	If not, why: Goals were achieved in part; at the beginning reduced consumption of lubricats; reduced costs of lubricants could be achieved; problems with the life-cycles of equipment, damages; causality still needs to be checked; HACCP was achieved; lubricants emissions could be reduced.				
3	Was the implementation of the new business model challenging in general?	①	❷	③	④
3 a	Difficulties during/with the implementation? Damages of equipment; changing of management responsibilities.				
4	Does the new business model provide cost reductions? Yes and no; cost of lubricants went down, but there is the problem with damaged equipments.	①	②	❸	④
5	Are you successful with the new business model?	①	②	❸	④
6	Were other companies involved in the implementation of this model (providers of equipment, disposal/recycling companies, etc.)?	YES		NO X	
	If yes, please specify:				
7	Characteristics of involved chemicals:				
	Value	①	❷	③	④
	Recyclability	①	②	❸	④
	Hazardousness	①	❷	③	④
8	Consequences of the model:				
	Product innovation	❶	②	③	④
	Earnings	①	②	③	④

<table>
<tr><td></td><td></td><td>Cost</td><td>①</td><td>②</td><td>③</td><td>❹</td></tr>
<tr><td></td><td></td><td>Supplier relationship</td><td>❶</td><td>②</td><td>③</td><td>④</td></tr>
<tr><td></td><td></td><td>Competition</td><td>①</td><td>❷</td><td>③</td><td>④</td></tr>
<tr><td></td><td></td><td>Mass of raw material/new product (also in %)</td><td>①</td><td>❷</td><td>③</td><td>④</td></tr>
<tr><td>9</td><td colspan="2">Is the model also suitable for very small enterprises (5 to 20 employees) or small product quantities?</td><td colspan="2">YES
X</td><td colspan="2">NO
O</td></tr>
<tr><td></td><td colspan="6">Reasons:
Low complexity</td></tr>
<tr><td>10</td><td colspan="6">Motives for the introduction of the new business model?
(See question 1)</td></tr>
<tr><td>11</td><td colspan="6">What are the main advantages/disadvantages of this model as compared to the traditional sales concept?

Higher productivity and reduced emissions.</td></tr>
<tr><td>12</td><td colspan="2">Which charging base (unit of payment) do you use?</td><td colspan="4">Pesos/t milled sugar cane</td></tr>
<tr><td>13</td><td colspan="2">Has this charging base any disadvantage from your point of view?</td><td colspan="2">YES
O</td><td colspan="2">NO
X</td></tr>
<tr><td></td><td colspan="6">If yes, please specify:</td></tr>
<tr><td>14</td><td colspan="2">What quantity of a new product could be saved?
kg, l, %, etc.</td><td colspan="4">50%</td></tr>
<tr><td>15</td><td colspan="2">What quantity of hazardous wastes could be reduced?
kg, l, %, etc.</td><td colspan="4">25%</td></tr>
<tr><td>16</td><td colspan="2">What was the decrease of emissions to the environment?
kg, l, %, etc.</td><td colspan="4">25%</td></tr>
<tr><td>17</td><td colspan="2">Would you implement the new business model once again?</td><td colspan="2">YES
X</td><td colspan="2">NO
O</td></tr>
<tr><td>18</td><td colspan="2">Will you use the business model also for other products?</td><td colspan="2">YES
X</td><td colspan="2">NO
O</td></tr>
<tr><td>19</td><td colspan="2">Will this model foster the implementation of REACH?</td><td colspan="2">YES</td><td colspan="2">NO</td></tr>
<tr><td></td><td colspan="6">Reasons:</td></tr>
<tr><td>20</td><td colspan="6">What would you change, complete or develop at this model?

Have competition between providers for the next period.</td></tr>
</table>

5.2.2 Electroplating

Elisa Arreola Valerio, Jorge Perez, Ignacio Sanchez, and Reinhard Joas

Case study information:

1. Information about the branch of industry and the conditions in the country

Based on the number of companies and jobs generated, the industry of electroplating in Mexico is an important sector in the economy of the country. The industry is located mainly in three areas: the metropolitan zone of the city of Mexico, Jalisco and Nuevo León.

Until June 1996 nationwide 794 registered establishments existed, 610 of which were classified as micro-companies, 138 as small companies, 29 as medium companies and 17 as large companies, based on the number of employees and the annual volume of sales. Of the total of establishments, 40% are located in the metropolitan zone of the city of Mexico, 28% in Jalisco, 19% in Nuevo León and 12% in the rest of the country.

Regarding the personnel employed, the electroplating industry generates 13,176 jobs, 10,312 of which are contributed by the sub-sector of the micro-small-medium companies, representing 78% of the total. The large companies contribute 2,864 jobs corresponding to 22%. The regional distribution of the personnel employed is 54% in the metropolitan zone of the city of Mexico, 20% in Jalisco, 18% in Nuevo León and 8% in the rest of the country.

As for the environmental aspects, this industry faces serious problems because most wastes and emissions generated are considered to be dangerous. In a study elaborated by the National Institute of Ecology in 1996, "Ciudades y giros prioritarios de acuerdo con su potencial contaminante", the electroplating ranks at the eighth place in the country regarding its contribution to the contamination through residual waters, and at the fifth place as far as the contamination through dangerous wastes concerns.

The most important emissions are residual waters deriving from the rinsing of pieces, exhausted process baths and from the draining originating during maintenance of the baths. An additional problem is represented by muds that are obtained in the processes of treatment of residual water, which must be confined given their toxic nature.

The industry of electroplating in Mexico has made diverse efforts to control its own emissions of polluting agents to the environment. Nevertheless, most of the companies of this industrial sector have serious difficulties, mainly due to the high costs of installation and operation of treatment systems, in addition to the high costs of final waste disposal.

2. Description of processes and services supplied by Chemical Leasing

In the following diagram the conventional nickel plate process that was carried out so far is shown. The process MARNIF A205 proposed by the service provider (indicated by the dashed line) implies a modification in the stage of nickel plate where the process MARNIF 205 was to be implemented (Fig. 1).

The new process started its operation in 2006. The Mexican Cleaner Production Centre evaluated the actual performance in order to check the economic and environmental benefits (reduction in the nickel consumption) and the functionality of the scheme of payment that forms the basis for the service of the chemical agents (brighteners) that was provided.

The project was successfully finished and a Chemical Leasing contract on the use of brightener was concluded (supported by the technical report). The unit of payment is Ampere-hours (US $ 0.14/10 Ampere-hour) corresponding to the surface of galvanised pieces. The consumption of nickel was reduced in the Chemical Leasing business model by 22% from 585 kg/year to 420 kg/year. Savings of around US $ 10,000/year could be reached, which depends highly on the increasing nickel price.

The previous results were included in a contract that was presented to both, the service provider and the service recipient and both accepted all clauses that were suggested for the document.

3. Tasks of service provider – tasks of service recipient

The service provider has the responsibility to adapt the operations in which its products are applied and to enable and give necessary technical support to the personnel of the user for the optimal operation of the process MARDI-INC.

Wherever applicable and according to the service provider's best judgement, the user may install a semiautomatic dispenser for the brighteners in the work tank.

For the implementation both companies bought an Ampere-hour equipment and installed a dosage for the brighteners in the electrolytic bath.

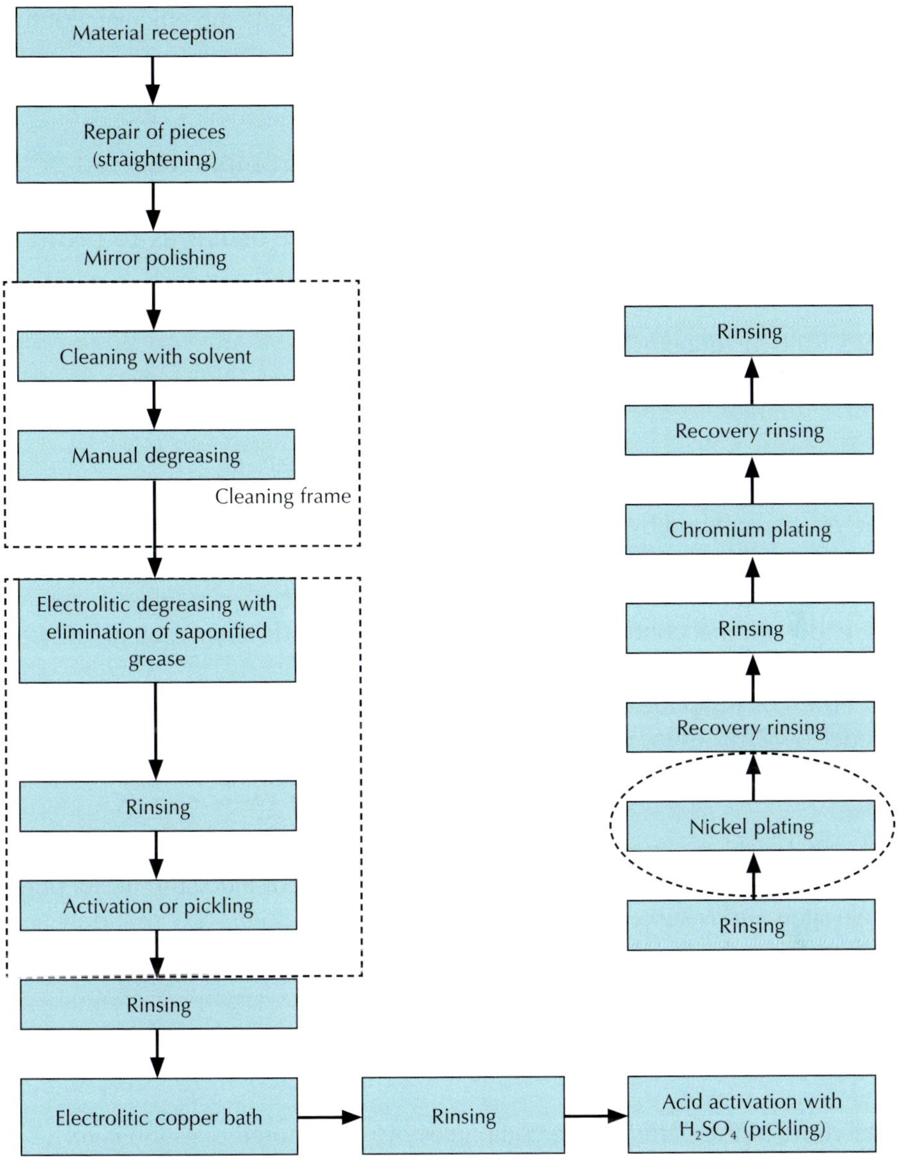

Fig. 1. Conventional nickel plating process

Both will offer the technical facilities to corroborate the efficiency of the process MARDIINC.

4. Functional unit as basis of payment (for example: cleaned or coated surface area, hours of operation, volume of purified water, etc.) and further agreements (in particular as to profit sharing and liability issues)

US $ 0.14/10 Ampere-hour.

5. Benefits

The process offered by the service provider allows to achieve a similar quality of the finished pieces with advantages in several aspects as compared to the conventional nickel plate process, as long as the bath is stabilised, operating under optimal conditions and using the polishers developed by the service provider.

The consumption of nickel was reduced in Chemical Leasing business model by 22% from 585 kg/year to 420 kg/year. Savings of around US $ 10,000/year could be achieved higly depending on the increasing nickel price.

The nickel price is the promoting factor for the decision to change the traditional process to a process like the one proposed. Since the nickel price is elevated and is subject to modifications according to the law of supply and demand, it is therefore a very volatile price.

6. Experiences

Interviews of the participating companies by a structured questionnaire.

Please find attached the questionnaires.

7. Involved partners

- Company name: Cromadora Delgado, S.A. de C.V. (CRODEL)
- Field of activity: Electroplating
- Role in the project: Service recipient

- Company name: MARDI Inc., S.A. de C.V. (MARDI)
- Field of activity: Electroplating
- Role in the project: Service provider

Fig. 2. Galvanising bath at Cromadora Delgado

Fig. 3. Metal pieces after treatment in galvanising bath

Questionnaire

Electroplating

Mardi (Supplier)

		① LOW	②	③	X HIGH
1	What were the key benefits expected leading to your participation in the project? **To offer a process option that presented advantages in environmental impact, quality and price to our client.**				
2	Were these goals achieved?	YES X		NO O	
	If not, why:				
3	Was the implementation of the new business model challenging in general?	①	❷	③	④
3	a Difficulties during/with the implementation? **The classic resistance to change on the part of the end users.**				
4	Is the new business model competitive	①	②	③	❹
5	Are you successful with the new business model?	①	❷	③	④
6	Could you use the experiences also for other products/offers?	①	②	③	❹
7	How many customers have accepted the new business model till now?	**1**			
8	Are these small (to 50 employees), medium-sized (51 to 250 employees) or big enterprises (more as 250 employees)?	**small**			
9	Is the application of chemicals a core competence of these enterprises?	YES X		NO O	
10	Were other companies involved in the implementation of this model (providers of equipment, disposal/recycling companies, etc.)?	YES O		NO X	
	If yes, please specify:				
11	Characteristics of involved chemicals:				
	Value	①	❷	③	④
	Recyclability	①	②	③	④
	Hazardousness	❶	②	③	④
12	Consequences of the model:				
	Product innovation	①	②	❸	④
	Earnings	①	②	③	❹
	Cost	❶	②	③	④
	Sales volume	❶	②	③	④
	Customer relationship	①	②	③	❹

	Competition	①	②	③	❹
	New customers	①	②	❸	④
	Mass of raw material/new product (also in %)	①	❷	③	④
13	Is the model also suitable for very small enterprises (5 to 20 employees) or small product quantities?	YES O		NO X	
	Reasons: **Resistance to change and administration problems.**				
14	Motives for the introduction of the new business model? • **reduce the environmental impact** • **reduce production costs** • **best control of raw materials (chemicals)** • **increase quality**				
15	What are the main advantages/disadvantages of this model as compared to the traditional sales concept? Advantages: **It is possible to establish a consumption average.** Disadvantages: **If many elements are to be considered, the method should be simplified.**				
16	Which charging base (unit of payment) do you use?	**(AH/HR) ($)**			
17	Has this charging base any disadvantage from your point of view?	YES X		NO O	
	If yes, please specify: **If many elements have to be considered, this factor cannot be correct for some of the parts, but during the implementation time user and supplier would adapt the price or bussiness index.**				
18	What quantity of a new product could be saved? kg, l, %, etc.				
19	What quantity of hazardous wastes could be reduced? kg, l, %, etc.	**60% by drag** **25% by deposition**			
20	What was the decrease of emissions to the environment? kg, l, %, etc.				
21	Would you implement the new business model once again?	YES X		NO O	
22	Will you use the business model also for other products?	YES O		NO X	
23	Will this model foster the implementation of REACH?	YES O		NO X	
	Reasons: **The model is not ideal for the process, because there are many complications during implementation.** **The conventional way of purchasing provides the same benefits, with the great advantage of facilitating acceptance.**				
24	What would you change, complete or develop at this model? **I would generate a new model in which sale is implemented as in a classic purchasing process. the primary target should be the reduction of the environmental impact with an economic benefit for the supplier as well as for the manufacturer. without clear economic benefit the quality that is offered to the final consumer is diminished.**				

Questionnaire

Electroplating

Cromadora Delgado (User)

			① LOW	②	③	X HIGH
1	What were the key benefits expected leading to your participation in the project? • reduction of consumption of dangerous raw materials • cost reduction of raw materials • increase the process control and quality					
2	Were these goals achieved?		YES X		NO O	
	If not, why:					
3	Was the implementation of the new business model challenging in general?		①	②	❸	④
3	a	Difficulties during/with the implementation? Strong opposition to the change of process.				
4	Does the new business model provide cost reductions?		①	②	❸	④
5	Are you successful with the new business model?		①	②	③	❹
6	Were other companies involved in the implementation of this model (providers of equipment, disposal/recycling companies, etc.)?		YES X		NO O	
	If yes, please specify: The supplier company and the technical support for new process implementation.					
7	Characteristics of involved chemicals:					
		Value	①	②	③	❹
		Recyclability	❶	②	③	④
		Hazardousness	①	②	③	❹
8	Consequences of the model:					
		Product innovation	①	②	③	❹
		Earnings	①	②	❸	④
		Cost	①	②	❸	④
		Supplier relationship	①	②	❸	④
		Competition	①	②	③	④
		Mass of raw material/new product (also in %)	①	②	❸	④

9	Is the model also suitable for very small enterprises (5 to 20 employees) or small product quantities?	YES X	NO O
	Reasons: Environmental and economic benefits.		
10	Motives for the introduction of the new business model? Ncrease the company competition and reduce process wastes.		
11	What are the main advantages/disadvantages of this model as compared to the traditional sales concept? Increased process control, pollution prevention, and economic earnings.		
12	Which charging base (unit of payment) do you use?	cost per Ampere hour	
13	Has this charging base any disadvantage from your point of view?	YES O	NO X
	If yes, please specify:		
14	What quantity of a new product could be saved? kg, l, %, etc.	22%	
15	What quantity of hazardous wastes could be reduced? kg, l, %, etc.	22%	
16	What was the decrease of emissions to the environment? kg, l, %, etc.	N/A	
17	Would you implement the new business model once again?	YES X	NO O
18	Will you use the business model also for other products?	YES X	NO O
19	Will this model foster the implementation of REACH?	YES X	NO O
	Reasons: Fortify the company image, market increase, raw material earnings and process control.		
20	What would you change, complete or develop at this model? More promotion.		

5.3 Chemical Leasing in Russia

5.3.1 Water Purification

Alexander Startsev and Rudolf Schott

Case study information:

1. Information about the branch of industry and the conditions in the country

A large (over 50%) part of glues, soaps and detergents in the Russian market is represented by either imported products or the production of Russian affiliations of foreign companies. The service provider in this case study is subsidiary of a worldwide known corporation and the largest producer of industrial and household glues, soaps and detergents in the North-Western region of Russia.

The service recipient is a share holding company specialising in R&D and service activities on industrial wastes disposal, wastewater purification, recycling of various chemicals and development of environment friendly regional programmes for hazardous wastes disposal and chemicals recycling. This small enterprise has a large experience in effective commercial implementation of efficient waste management approaches (processing, disposal, recycling) in different branches of industry such as galvanic processes, metallurgy, and production of organic compounds. The company started its R&D and commercial implementation activities in the late 1980s. In the recent years they are successfully performing a series of innovative projects relating to wastewater purification in different branches of industry. A high potential and prospects of these projects are determined by a rapid growth of the Russian economy, particularly chemical branches, causing a significant impact on the environment, outdated and poor conditioned wastewater purification facilities failing to decontaminate the effluents to the required level and strict environment protection legislation adopted in the country.

2. Description of processes and services supplied by Chemical Leasing

Before the application of Chemical Leasing (ChL) the service company supplied chemicals and materials (iron chloride $FeCl_3$, sodium hydroxide NaOH, filter packs) for wastewater purification. However, the level of water decontamination at the existing facilities was insufficient in view of the glue production increase. The implementation of the ChL business model based on payment for the amount of purified water motivated both companies to cooperate in their efforts for process modernisation. The reconstruction of wastewater processing facilities and application of an advanced decontamination technology according to a know-how suggested by the service company provided the increase of purification quality to an environmentally appropriate level and a significant reduction in the consumption of all the chemicals used for wastewater decontamination.

3. Tasks of service provider – tasks of service recipient

The service provider has developed and installed the improved wastewater purification process at the user's facilities and currently supplies the minimised amounts of required chemicals ($FeCl_3$, NaOH) on the basis of Chemical Leasing. The company also takes care of the user's personnel training in the new process operation, the residual solid wastes transportation and disposal.

The service recipient has performed all up-scaling activities required for the improved wastewater process installation and currently effectively performs wastewater purification using the chemicals supplied on the basis of Chemical Leasing.

The North-Western International Cleaner Production Centre (NWICPC) as the project coordinator organised the communication (meetings, workshops, agreement preparation) between the representatives of the service provider and recipient, provided them with the information about the advantages of the Chemical Leasing business model and promoted the national and international support of the project.

4. Functional unit as basis of payment (for example: cleaned or coated surface area, hours of operation, volume of purified water, etc.) and further agreements (in particular as to profit sharing and liability issues)

Payment according to the Chemical Leasing agreement between the parties is based on the amount (cubic meters) of purified water. The process optimisation provided the reduction of 1 m^3 wastewater purification cost from the initial value of about 27 Euro to 13 Euro established according to the Chemical Leasing agreement.

5. Benefits

Economic benefits: More than 50% decrease of wastewater purification cost and required chemicals consumption mutually profitable for participating companies.
Quality and organisational benefits: The implementation of the advanced wastewater purification process is profitable for the service provider only in the case of Chemical Leasing application since the amount of supplied chemicals is significantly reduced.
Environmental benefits: The installation of the new wastewater purification process provided the reduction of impurities contents in the glue production effluents to a level below the sanitary standards. Furthermore, the amount of environmentally hazardous chemicals ($FeCl3$, NaOH) used for wastewater purification is significantly reduced.

6. Experiences

(Interviews of the participating companies by a structured questionnaire.)

In the interview relating to the experience of Chemical Leasing application the General Director of the service provider underlined that the new business model was beneficial for the company in respect of economic profit and opportunity for a most effective introduction of innovations and know-how. The Deputy Technological Manager of the company using the service indicated that the new business relationship was economically beneficial and provided the fastest and mostly efficient approach to solve environmental problems that the company had been facing before.

7. Involved partners and country information

Supplier

- Company name: ERG
- Field of activity: R&D and service company specialising in the development of waste purification and disposal processes
- Role in the project: Service provider of industrial wastewater purification including the supply of chemicals, equipment, personnel training and residual wastes disposal
- Location: Russia, St-Petersburg, Torzhkovskaya str. 4, office 413
- Contact person: Dr. Sergey Karpov, General Director

User

- Company name: Henkel-ERA
- Field of activity: Production of industrial, constructing and household glues, synthetic soaps and detergents
- Role in the project: User (service recipient) of chemicals and equipment for wastewater purification
- Location: Russia, 187000, Tosno, Leningrad region, Moskovskoe shosse, 1
- Contact person: Dr. Wolfgang Bürger, Production Manager
- Website: www.henkel.com, ru.henkel.com

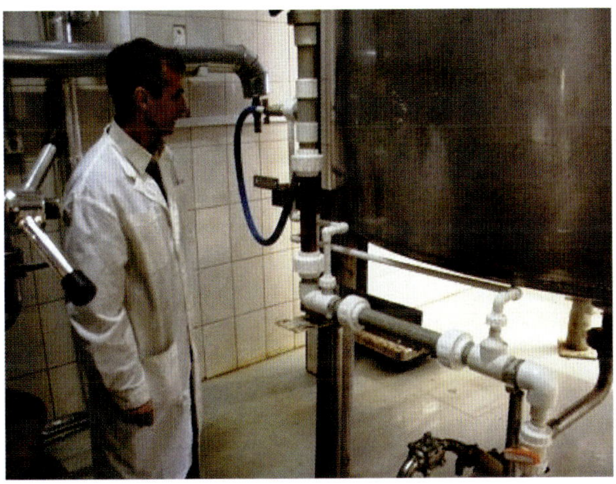

Fig. 1. Waste water treatment at Henkel

Fig. 2. Production of glues at Henkel

Questionnaire

Water Purification

ERG (Supplier)

		① LOW	②	③	④ HIGH
1	What were the key benefits expected leading to your participation in the project? 1. **implementation of new equipment, chemicals and process for the purification of glue production wastewater** 2. **more than 50% decrease of wastewater purification cost and required chemicals consumption mutually profitable for participating companies**				
2	Were these goals achieved?	YES X		NO O	
	If not, why:				
3	Was the implementation of the new business model challenging in general?	①	②	❸	④
3 a	Difficulties during/with the implementation? **Long period of optimising the process.**				
4	Is the new business model competitive	①	②	③	❹
5	Are you successful with the new business model?	①	②	③	❹
6	Could you use the experiences also for other products/offers?	①	②	③	❹
7	How many customers have accepted the new business model till now?	**about 5**			
8	Are these small (to 50 employees), medium-sized (51 to 250 employees) or big enterprises (more as 250 employees)?	**small and medium-sized**			
9	Is the application of chemicals a core competence of these enterprises?	YES O		NO X	
10	Were other companies involved in the implementation of this model (providers of equipment, disposal/recycling companies, etc.)?	YES X		NO O	
	If yes, please specify: **Krasny Bor.**				
11	Characteristics of involved chemicals:				
	Value	①	②	❸	④
	Recyclability	❶	②	③	④
	Hazardousness	①	❷	③	④
12	Consequences of the model:				
	Product innovation	①	②	③	❹
	Earnings	❶	②	③	④

		Cost	①	❷	③	④
		Sales volume	①	②	❸	④
		Customer relationship	①	②	❸	④
		Competition	①	②	③	❹
		New customers	①	②	❸	④
		Mass of raw material/new product 100%	①	②	③	❹

13	Is the model also suitable for very small enterprises (5 to 20 employees) or small product quantities?	YES X	NO O
	Reasons: The application of Chemical Leasing business model provided more than 50% decrease of wastewater purification cost and required chemicals consumption thus providing a mutual profit for both participating companies. Small enterprises also can collect wastes for cyclic purification (for example once per week).		
14	Motives for the introduction of the new business model? Implementation of new technologies in the entire plant.		
15	What are the main advantages/disadvantages of this model as compared to the traditional sales concept? Opportunity for establishing mutually profitable long-time relations.		
16	Which charging base (unit of payment) do you use?	cubic meter	
17	Has this charging base any disadvantage from your point of view?	YES O	NO X
	If yes, please specify:		
18	What quantity of a new product could be saved? kg, l, %, etc.	about 50% of reagents can be saved	
19	What quantity of hazardous wastes could be reduced? kg, l, %, etc.	concentration of the hazardous wastes will be less than the maximum permissible concentration established by law	
20	What was the decrease of emissions to the environment? kg, l, %, etc.	organic compounds up to 98%	
21	Would you implement the new business model once again?	YES X	NO O
22	Will you use the business model also for other products?	YES X	NO O
23	Will this model foster the implementation of REACH?	YES X	NO O
	Reasons: Ability to control production process and life cycle of the substance.		
24	What would you change, complete or develop at this model? Implementation of this model in all treatment stations of the plant.		

Questionnaire

Water Purification

Henkel-ERA (User)

		① LOW	②	③	④ HIGH
1	What were the key benefits expected leading to participation in the project? 1. implementation of new equipment, chemicals and process for the purification of glue production wastewater 2. more than 50% decrease of wastewater purification cost and required chemicals consumption mutually profitable for participating companies				
2	Were these goals achieved?	YES X		NO O	
	If not, why:				
3	Was the implementation of the new business model challenging in general?	①	②	❸	④
3 a	Difficulties during/with the implementation? Long period for optimisation of process.				
4	Does the new business model provide cost reductions?	①	②	③	❹
5	Are you successful with the new business model?	①	②	③	❹
6	Were other companies involved in the implementation of this model (providers of equipment, disposal/recycling companies, etc.)?	YES O		NO X	
	If yes, please specify:				
7	Characteristics of involved chemicals:				
	Value	①	❷	③	④
	Recyclability	❶	②	③	④
	Hazardousness	❶	②	③	④
8	Consequences of the model:				
	Product innovation	①	②	③	❹
	Earnings	❶	②	③	④
	Cost	①	❷	③	④
	Supplier relationship	①	②	③	❹
	Competition	①	②	❸	④
	Mass of raw material/new product 100%	①	②	③	❹

9	Is the model also suitable for very small enterprises (5 to 20 employees) or small product quantities?	YES O	NO X
	Reasons: The application of the Chemical Leasing business model provided 50% decrease of wastewater purification cost and required chemicals consumption thus providing a mutual profit for both participating companies. Small enterprises also can collect wastes for cyclic purification (for example once a week).		
10	Motives for the introduction of the new business model? Implementing new technologies in other purification or production processes.		
11	What are the main advantages/disadvantages of this model as compared to the traditional sales concept? Opportunity for establishing mutually profitable long-time relations and replacement of old technologies using innovations.		
12	Which charging base (unit of payment) do you use?	cubic meter	
13	Has this charging base any disadvantage from your point of view?	YES O	NO X
	If yes, please specify:		
14	What quantity of a new product could be saved? kg, l, %, etc.	about 50% of reagents can be saved	
15	What quantity of hazardous wastes could be reduced? kg, l, %, etc.	concentration of the hazardous wastes will be less than the maximum permissible concentration established by law	
16	What was the decrease of emissions to the environment? kg, l, %, etc.	organic compounds up to 98%	
17	Would you implement the new business model once again?	YES X	NO O
18	Will you use the business model also for other products?	YES X	NO O
19	Will this model foster the implementation of REACH?	YES X	NO O
	Reasons: User will use product of proper quality.		
20	What would you change, complete or develop in this model? Implementation of this model in the entire plant will allow us to reduce production costs and impact purification facilities. It will also be our contribution to environmental protection.		

Chapter 6

Third-Party Quality Assurance and Certification Chemical Leasing: Optimisation by Certification

Ulrich Nagel and Peter Schaff

Chemical Leasing is a business model benefiting not only lessor and lessee, but also the environment. Quality management is one of the key factors in ensuring that all stakeholders will actually be able to realise the benefits involved in Chemical Leasing, and is of particular importance at the interfaces between leasing partners. TÜV SÜD Management Service GmbH, a TÜV SÜD company, has decades of experience in the certification of various management systems throughout all industries. This contribution shows that "Certified Chemical Leasing", a standard developed by TÜV SÜD, ideally integrates quality, environmental and occupational health and safety elements with specific requirements of the chemical industry. The key is integrated management systems! The article also reveals, however, that even assessment of individual segments – quality, environmental or occupational health and safety management – strengthens the Chemical Leasing process to the advantage of all stakeholders and the environment.

Why quality management?

Chemical Leasing, as is generally known, aims at the benefit-oriented and thus more efficient use of chemicals to realise both ecological and economic advantages. Quality assurance by an impartial third party assists interested parties to secure the benefits aimed at through Chemical Leasing, can initiate and promote Continuous Improvement Processes in participating companies and prevents misuse, which could discredit the model and thus jeopardise existing opportunities. A well-established quality assurance system can also strengthen the basis of trust between stakeholders, increase acceptance of Chemical Leasing among third parties, e.g. governmental organisations, sponsors, investors and, last but not least, the staff involved, and widen existing interest in the model.

Focus on quality, environmental protection and occupational health and safety

The interfaces between Chemical Leasing partners are of paramount importance for quality assurance in this context. Owing to the highly heterogeneous structure of the participating companies, ranging from small enterprises in developing countries to chemical multinationals, a graduated quality assurance system will be needed. Within the scope of individual certification processes, TÜV SÜD Management Service uses criteria catalogues, which are flexibly customised to on-site conditions in the individual company. With respect to quality, environmental and occupational health and safety management, in Chemical Leasing, the term "quality assurance" needs to be defined more broadly. It applies to the quality of performance in technical, occupational health and safety as well as environmental areas. In particular, quality assurance also regulates roles and responsibilities for the quality of various services and any liability issues arising thereof, thus also ensuring legal compliance and effectiveness.

Quality assurance principles

A quality assurance programme for Chemical Leasing projects has to be based on the following principles:

1. The certificate is awarded following assessment by an independent third-party organisation.
2. The project must involve verified ecological benefits.
3. Clear assignment of responsibilities to stakeholders, in accordance with risk management, must be ensured.
4. National regulations are observed.
5. National and international standards, where applicable and suitable, must be consulted.

Minimum requirements for Chemical Leasing in terms of quality assurance include:

- State-of-the-art and statutory requirements plus specific requirements defined by the chemical industry applicable to the substances used have been fulfilled.
- Safe and ecological operation of installations and systems is ensured.
- Personnel are protected from safety and health hazards in the workplace.

- Stakeholders' (lessor's and lessee's) interests are considered on a fair basis characterised by mutual trust.

Quality assurance verifies that

- requirements for state-of-the-art technical equipment are satisfied,
- personnel training and qualification criteria are fulfilled,
- process and product safety is ensured,
- ecological standards and applicable limits are observed as a minimum requirement,
- a suitable risk management system has been established to address potential emergencies or failures,
- the Continuous Improvement Process – the core of every quality management system – is effective and proactively applied.

Ideally, Certified Chemical Leasing will ensure expedient and target-group specific bundling of quality, environmental and occupational health and safety management aspects plus fulfilment of the specific requirements of the chemical industry (Fig. 1).

Figure 1

The four building blocks in detail

Quality management

ISO 9001, the classic among international quality management standards, describes a quality management approach for the entire organisation. Striving for continuous improvement, quality management focuses on legitimate customer requirements – here the keyword is Continuous Improvement Process (CIP). An accredited third-party certification body – for example TÜV SÜD Management Service – can confirm that the organisation has established a quality management system, maintains the required documentation and actually implements all measures needed to achieve quality objectives. A functioning quality management system establishes a basis for an effective and efficient partnership in Chemical Leasing. Nevertheless, certification of the quality management system is not a must requirement for Certified Chemical Leasing. However, when conducting certification according to the "Certified Chemical Leasing" standard, TÜV SÜD experts also include key elements of ISO 9001 which are crucial for the success of the leasing partnership.

Organisations consistently integrating quality management into their culture enjoy a host of benefits not limited to Chemical Leasing itself. Quality management ensures standardised processes and consistently high product and service quality – independent of how strong the individual players are going on a specific day. It helps to respond quickly and flexibly to new general legal or technical conditions and changing customer requirements. Quality management enables multinationals to adapt to specific national requirements while maintaining a high level of product and service quality across all locations and borders.

Environmental management

Environmental management integrates cost-effectiveness, customer focus and ecological responsibility in enterprises. Management systems also facilitate successful implementation of the organisation's environmental policy and statutory requirements. The ISO 14000 series of standards offers an appropriately comprehensive and systematic approach, which ensures that environmental objectives and targets will be considered throughout all areas during the planning, implementation and control of corporate activities. An environmental management system also helps to define, initiate and optimise the environmental effectiveness of processes. Ongoing comparison between actual status and statutory requirements is another element firmly anchored in the system. Environmental management systems may speed up approval procedures and allow potential resource savings to be systematically realised. Another benefit of environmental management systems is that they reduce the risk of violation of environmental regulations – the keyword here being legal compliance. Moreover, liability risks also become calculable.

The following applies to ISO 14001 and other environmental management standards: Certification is not a must requirement for "Certified Chemical Leasing". However, TÜV SÜD's criteria catalogues for Chemical Leasing also include key elements of the ISO 14001 standard.

Occupational health and safety management

Occupational health and safety management is an important subject, in particular in conditions where chemicals are handled. High standards of occupational health and safety strengthen employee motivation, resulting in lower absenteeism rates and fewer production stoppages. When handling chemicals, in particular, potential physical and psychological stress must be taken into account. Organisations have the duty of preserving the physical, mental and psychological wellbeing of their personnel while permanently sustaining and enhancing their work performance. If top management and on-site executives regularly address occupational health and safety, unnecessary costs caused by accidents will be avoided. The guidelines issued by the International Labour Organisation (ILO) and the OHSAS 18001 (Occupational Health and Safety Assessment Series) or SCC (Safety Certificate Contractors) standards are some of the documents ensuring a systematic approach.

Occupational health and safety also become particularly important in view of the increasing tendency towards fewer and fewer statutory regulations, resulting in a higher degree of freedom, but also in a significantly higher level of responsibility for entrepreneurs and executive staff. Important keywords in this context, also in terms of Chemical Leasing, include: harmonisation, liberalisation and deregulation. An occupational health and safety management system helps to make the most of this freedom while still taking full account of the responsibilities involved.

The following also applies to systematic occupational health and safety management: Companies not only benefit within the scope of Chemical Leasing. OHSAS 18001 and other occupational health and safety standards, for example, support organisations in identifying and eliminating hazard factors. Informing employees openly about prevention goals adds to their motivation to become proactively involved in the achievement of these goals. Application of a management system allows occupational health and safety performance to be compared on national and international levels, in both industrialised and newly industrialised countries.

Last but not least, the following also applies to occupational health and safety management: When certifying Chemical Leasing projects, TÜV SÜD specialists do not require complete compliance with an occupational health and safety standard, e.g. OHSAS 18001, although the key elements of an occupational health and safety system must be observed.

Specific requirements of the chemical industry

Special process and product specifications have been set forth for companies in the chemical industry. Additionally, specific limits must be observed for pollutant or VOC (volatile organic compounds) emissions. Specific maximum allowable concentrations at workplaces must not be exceeded. Both the transport of chemicals and the recycling or disposal of chemical wastes are subject to particular requirements. Periodic hazard assessments – which in Germany, for example, are mandatory according to the Ordinance on Industrial Safety and Health (Betriebssicherheitsverordnung, BetrSichV) – are a must requirement in the chemical industry. To achieve a Continuous Improvement Process, product and process requirements must be identified, evaluated and monitored on a regular basis.

TÜV SÜD's expertise

Using techniques and technologies to benefit people, improving them continuously and keeping people and the environment safe. This has been the corporate mission of TÜV SÜD for over 140 years. The corporate objective set forth in the TÜV statutes at its foundation was "To protect people, the environment and property against the adverse effects of technology". However, as early as in the 19th century the company extended its scope beyond this objective, and also issued recommendations on cost-effective plant operation. Soon, the basic concerns that have driven TÜV SÜD operations since that time emerged: Safety, cost-effectiveness and quality assurance – three goals that also apply to the "Certified Chemical Leasing" model.

For over 25 years, TÜV SÜD Management Service GmbH has been dedicated to the certification of quality and other types of management systems. The experts at TÜV SÜD Management Service are familiar with all relevant international and national standards and, working in interdisciplinary teams, have also developed their own standards. "Certified Chemical Leasing" is also a standard tailored to practical needs and developed by TÜV SÜD Management Service, which has decades of experience in the chemical industry throughout all major economic regions in Europe, America and Asia and profound expertise in quality, environmental and occupational health and safety management.

Longstanding and international experience, interdisciplinary approach and impartiality are TÜV SÜD attributes, which also benefit companies requesting assessment and certification in the field of Chemical Leasing.

Certification process

How does a certification process according to the "Certified Chemical Leasing" standard work? TÜV SÜD Management Service does not take action on its own account. All partners in a Chemical Leasing project can commission TÜV SÜD to carry out certification, e.g. chemicals producers, process equipment manufacturers, users of chemicals, disposal and recycling companies and many more (Fig. 2).

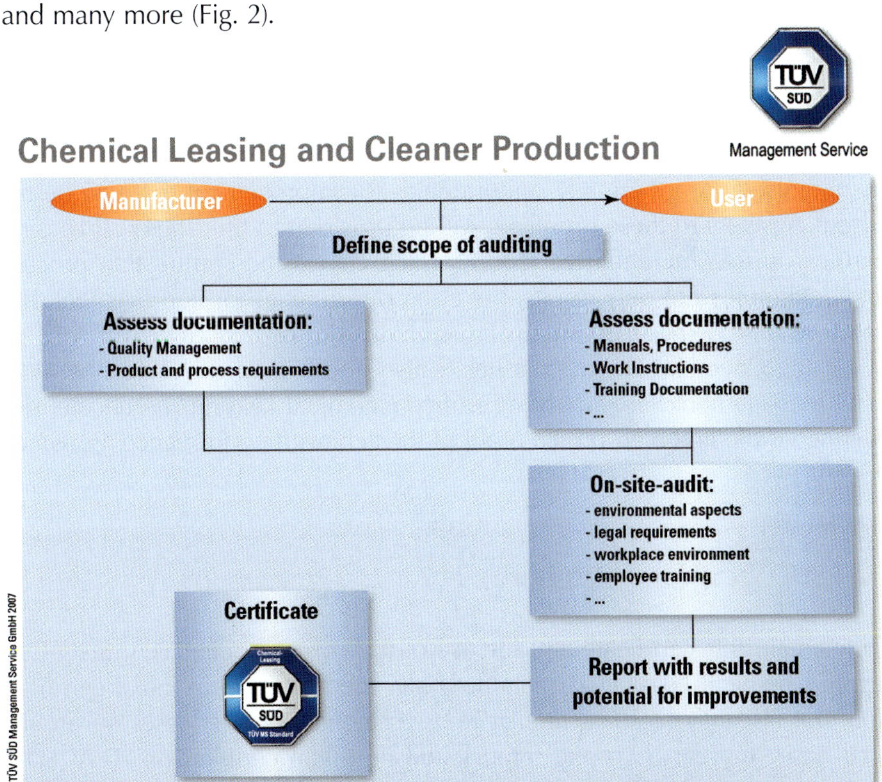

Figure 2

Before the actual start of the process, the scope of certification is defined jointly by the Chemical Leasing partners and TÜV SÜD, taking the stakeholders' interests and capabilities into account while considering indispensable requirements of quality, environmental and occupational health and safety management. Subsequently, the certification process and schedule are laid down in detail. The first step in the actual certification process is the review of relevant documents, comprising firstly the contractual obligations of the Chemical Leasing partners, and secondly documented procedures and work instructions established for the on-site production process. The four-pronged approach, involving quality, environmental and occupational health and safety management plus the specific requirements of the chemical industry, plays a major role in both document review carried out by auditors and on-site assessment, the "certification audit". In document review, TÜV SÜD specialists focus on aspects including commitment to quality management, product and process requirements, manuals and documented procedures, work instructions or training plans. During the on-site audit, experts verify on the spot, on the company's premises, whether quality and environmental requirements have been complied with, assess how legal requirements are implemented in practice, how individual workplaces are designed, how recycling and disposal, for example, are regulated, and so on.

Does the organisation comply with the agreed requirements? What problems and nonconformities have been revealed? How can cost-effectiveness be optimised? Are there any opportunities for improvement that can be realised? Answers to these questions will be provided to the Chemical Leasing partners in a comprehensive report, which covers the certification process and documents not only results, but also opportunities for improvement, the keyword here being "Continuous Improvement Process".

If the Chemical Leasing partners satisfy the criteria developed by TÜV SÜD Management Service, the "Certified Chemical Leasing" certificate, delivering visible and high-profile proof of the achieved performance, is awarded.

TÜV SÜD Octagon – Pro Tem

To maintain this certificate and the use of the corresponding TÜV SÜD Octagon, Chemical Leasing partners not only need to undergo initial certification, but also have to pass regular re-assessments, in most cases on an annual basis. In years two and three following initial certification, TÜV SÜD experts apply a sampling approach to verify that the organisation continues to comply with requirements. In year four, the entire certification process is repeated within the scope of "re-certification". In extreme cases, if major nonconformities with requirements are revealed in a company, certificates may be withdrawn immediately. The positive side of this coin: Organisations holding a "Certified Chemical Leasing" certificate and thus being authorised to use the corresponding TÜV SÜD Octagon document that they participate in a Chemical Leasing project at an advanced level. The time limit and the possibility of immediate intervention in cases involving violation and misuse emphasise the trustworthiness of the certification and the corresponding logo.

Organisations that have already implemented an integrated management system, for example as per ISO 9001, ISO 14001 and OHSAS 18001, may benefit from potential synergies. Other companies which are motivated by Chemical Leasing to start addressing the relevant standards will benefit from their implementation throughout their entire organisation – not only within the context of Chemical Leasing.

Criteria catalogue

Organisations striving to gain certification according to the TÜV SÜD "Certified Chemical Leasing" standard must fulfil a host of criteria in both theory and practice – ranging from energy management in production, reduction of emissions to air, water and soil, safe workplaces, regular measurement of key parameters, and waste and waste-water disposal to the selection of suitable means of transport (Fig. 3).

Within the scope of their assessment, or audit, experts ask questions, including:

- Does the company take steps to ensure sustainable energy use? What instruments does it employ to do so?
- How does the organisation identify, document and implement information about relevant environmental requirements?
- What measures does the company take to organise and ensure observance of emission limits at individual workplaces?
- How is disposal of recyclables and waste organised and how is it realised in practice?
- Does the organisation proactively address "critical" situations? What provisions does it make and what drills and practices does it perform?
- How does the organisation translate environmental requirements into instructions for the individual workplace?
- How do energy supply, waste and waste-water disposal function in the individual production and work units?

Chemical Leasing Certification
Environmental aspects check list

TÜV SÜD Management Service

Requirement / subject	Docu-mentation	E	Requirements implemented / documented in	Audit notes	E
Procedures to identify the environmental aspects of: - activities, - products, - services, which have or can have significant impacts on the environment.			o P: Identification of environmental aspects o Environmental review, where appropriate o Evaluation system for prioritizing environmental aspects o Evaluation system for prioritizing calls for action o ABC analysis (e.g. action required urgently, in the medium term, not at all),		
Factors to be considered: – *waste management* – *contamination of land* – *releases to water* – *emission to air* – *use of raw materials and natural resources* – *use of energy* – *current and relevant-past activities, products and services* – new or modified activities, products and services; – planned and new developments – *normal and abnormal operating conditions,* – *start-up and shut-down conditions,* – *emergency situations,* – *availability of reliable data,* – the possibilities of monitoring environmental aspects, – the possibilities of controlling and influencing environmental aspects.			o List of significant environmental aspects o Environmental review report, where appropriate o Register of plants and/or machinery o List of environmentally-relevant plants o Input/output balance sheet o Lists of raw, operating and auxiliary substances, by-products, intermediate and final products o Energy consumption o Waste balance sheet, waste management strategy o Waste-water analysis report o Report on waste-water monitoring o Report on the examination of contaminated sites o Plan of emissions sources o Noise-exposure map o Emissions measuring report o Hazardous-substances register		
Ensuring that significant environmental aspects are taken into account in Chemical Leasing.			o Action plans o Environmental objectives and targets o Documented procedures for operation control		

Figure 3

- What exhaust-air systems are available, do they function correctly and are they regularly serviced and maintained?
- According to which criteria does the organisation select means of transport for work equipment and products? Is resource-saving technology employed in transport?
- Does the organisation maintain a hazardous substances register? Is this register consistently updated and is safe handling of these substances implemented in work processes?
- Is the organisation committed to replacing hazardous substances with environmentally friendly ones?
- Are relevant explosion protection regulations observed in areas where chemicals are stored?
- Are sources of emission systematically identified and documented? Does the organisation have its own strategy, e.g. aimed at reducing CO2 emissions?
- Does the organisation seriously address the subject of noise emissions?
- Does the organisation maintain valid training certificates proving that personnel are trained regularly in the handling of hazardous materials?

Objective evidence to answer these questions is sought in both management system documentation and on-site interviews conducted with executives and employees. Evidence is documented and evaluated. On the basis of this evidence and these evaluations, an overall evaluation is then arrived at, which results in certificate issue and demonstrates existing opportunities for improvement.

Practical examples

Various patterns of Chemical Leasing including the accompanying quality management certification are possible, for example in case two companies agree on a Chemical Leasing project. Or in case a chemicals producer launches a cooperation with a process equipment manufacturer to provide optimised services to the Chemical Leasing partner. Another possibility is a joint venture between various partners, e.g. chemicals producers, process equipment manufacturers, transport and disposal companies, and a manufacturer planning to use Chemical Leasing services.

Examples for such applications in practice have been explained in the previous chapters.

Within the scope of certification, TÜV SÜD Management Service verifies on site, at the premises of all parties involved, whether all criteria set forth in the Chemical Leasing agreement and all indispensable state-of-the-art requirements for quality, environmental and occupational health and safety management have been implemented and complied with. Initial pilot projects which will confirm whether TÜV SÜD's "Certified Chemical Leasing" standard is fit for practice and actually adds value to stakeholders have been planned in cooperation with the United Nations Industrial Development Organisation (UNIDO). Given this, Chemical Leasing certificates and the TÜV SÜD Octagon for Chemical Leasing projects are expected in the near future.

Conclusions

Quality assurance is a must task to realise the opportunities offered by the Chemical Leasing model and prevent misuse and fraud. Certification by an impartial third party, e.g. TÜV SÜD Management Service, ensures that Chemical Leasing will actually generate sustainable solutions, that relevant criteria of quality, environmental and occupational health and safety management will actually take effect and that the specific requirements of the chemical industry will be observed. Based on decades of experience in the chemical industry and in quality management, TÜV SÜD Management Service has developed the "Certified Chemical Leasing" standard, which pursues an interdisciplinary approach. The TÜV SÜD Chemical Leasing certificate plus the accompanying Octagon strengthen confidence both in certified organisations and the Chemical Leasing model.

Chapter 7

Chemical Leasing – Legal Questions

Michael Wittmann

In the course of the implementation of the Chemical Leasing business model not only the technical and the economic feasibility, but also the general legal conditions have to be clarified. In this context questions in the fields of private law and administrative law arise.

1. The most important questions concerning private law are: What kind of contract is needed to implement a Chemical Leasing business model, what is the difference between such a contract and a purchase contract regulating a conventional business model, and what provisions should be included in the contract?
2. Questions concerning administrative law: Which obligations under administrative law have to be complied with and who is responsible for their compliance? What are the differences compared to the conventional model? The most important administrative matters in this context concern trade law – in particular the industrial plants law and the waste management law.

Below these questions are dealt with on the basis of the Austrian legal system.

Ad 1: Contracts

According to the Austrian law of contract there is a principle of freedom of contract. The contracting parties can freely agree on mutual obligations, only the most important contracts (purchase, tenancy, donation, etc.) are standardised by law. There is no obligation of standardisation of contracts. Therefore new business models, such as leasing or franchising, can be established in economic relationships, irrespective of the contracts standardised by law. In case of any legal dispute due to such an atypical contractual relationship, jurisdiction is first of all based on court rulings regarding the type of contract regulated by law which is most similar to the respective atypical contractual relationship. The court rulings on leasing contracts, for example, have developed according to the judicature on purchase and tenancy, since the leasing business is to be considered as a mixture between purchase and tenancy. When a new business model is introduced, it is thus of great interest from the legal point of view, what kind of contract is used and which of the established contracts is most similar to the new business.

The traditional model of supplying chemicals is based on the purchase contract. The supplier sells to the user a certain quantity of the chemical substance at a certain price. (This entails an incentive which is unfavourable from the environmental point of view: The supplier is the more successful, the more chemicals it sells. This should be avoided by applying the new business model.)

There is no doubt about the fact that the purchase contract is the most important contract in economic relationships, it is strictly regulated by the Austrian General Civil Code (AGBG Allgemeines Bürgerliches Gesetzbuch), the main code of Austrian private law. There the obligations of buyer and seller are defined, there are also rules for the most important potential causes of conflict and there is an abundance of court rulings. In short: The purchase contract is a routine business operation and there is an abundance of legal reference material.

Regarding Chemical Leasing the situation is quite different: It is not mentioned at all in the Austrian General Civil Code, it is an atypical contract which is yet to be developed by lawyers or legal divisions of enterprises planning to do business according to the Chemical Leasing model. As mentioned above it is important in this context to find out which of the established types of contract is most similar to the new contractual relationship.

There is comprehensive jurisdiction available regarding the leasing business commonly used in car trade. However, this business type cannot serve as a model for Chemical Leasing. Although discussions about a desired service-oriented and ecologically reasonable chemicals management business, the so-called Chemical Leasing, have been going on for years already, any similarity to the well-known leasing contract is rather superficial. If you need a car, you need not buy it, you can just lease it and get all benefits of having a car. The enterprise which needs a chemical substance need not buy it either, if it opts (together with its supplier) in favour of Chemical Leasing, and it will get the benefits of this chemical substance. However, these are all similarities, while the differences are much more prevailing. The car which you lease will become your own car after some time. However, in Chemical Leasing the user will never (if at all only temporarily) acquire ownership of the chemical substance to be used.

Tenancy cannot serve as a model either, as the business objective is not to sell a certain quantity of a chemical substance, which is rented so to speak. This would lead to the same undesired incentive as in case of a purchase: It would mean that business success depends on the quantity of the chemical substance used. Instead the business goal in Chemical Leasing is the benefit of the chemical substance, which can be considered to be a service.

It thus seems to be an atypical service contract: The former supplier, now lessor, undertakes to make the benefit of the chemical substance available to the user (lessee), and to take over, if required, certain further services in connection with the use of this chemical substance.

This atypical service contract is most similar to a contract for works and services, as a certain result is owed by the debtor like in a contract for works and services. A typical example for a common contract for works and services is commissioning a master builder to build a house on a plot of land. However, in Chemical Leasing the contract cannot be fulfilled by delivery and acceptance by the factory. The house will be completed sometime. However, chemicals will be permanently needed by the enterprises using them. Therefore the contract has to be de devised as a long-term obligation, and in this respect it is similar to the service contract or the tenancy. Also in case of more comprehensive services on the part of the supplier a service contract or tenancy can serve as a model.

The content of the contract will depend decisively on the desired level of integration, which means how many service tasks will be assumed by the supplier in connection with chemicals management. If a high level of integration is desired, supplier and user can found a company for the purpose of chemicals management and regulate the mutual obligations in a company agreement.

Otherwise it will be an atypical service contract, which is most similar to a contract for works and services, containing, however, elements of a service contract and tenancy.

It is important to include the following clarifying points in such a contract:

The definition of result owed by the debtor: What is the "benefit of a chemical substance" and how can be proven that the supplier has made available the benefit to the user. For this definition exact knowledge about the plant and the process in which the chemical substance is used is required.

The agreed fee and the factor on the basis of which it is charged: This could typically be a certain period of time or a certain production volume.

The obligations and the fields of responsibility of the contracting partners should be described as precisely as possible. This, again, predisposes an exact knowledge of the specific conditions of the company.

Rules on the most important cases of conflict in question should be provided for in the contract: warranty provisions, contractual penalties, etc.

To summarise it one could say that the Chemical Leasing contract will be an atypical service contract which is most similar to a contract for works and services, but contains also elements of service contract and tenancy. In the case of high integration and foundation of a company for the purpose of chemical management the mutual obligations can be regulated on the basis of a company agreement.

There cannot be a "sample contract" on Chemical Leasing because due to great variety within the framework of this business model the necessary consideration of company-specific conditions would not be possible.

Ad 2: Administrative law

In connection with Chemical Leasing the industrial plants law of the trade regulations and the waste management law are the most important matters under administrative law.

As opposed to private law it is clearly defined in the legal provisions under administrative law who is obliged to behave in which way. Usually there is no room of manoeuvre for free agreements. If for example an ordinance under industrial plants law stipulates an emission threshold value it will have to be complied with and the owner of the plant will be responsible for the compliance with the threshold value.

The legal form of the business activity cannot change anything regarding the threshold value and its binding character. However, with a Chemical Leasing with a high level of integration the person responsible can change. It could, for example, turn out to be useful that the supplier rents the plant of the user (or the parts of the plant where the chemical substance is used), and becomes thus the tenant of the industrial plant and responsible for the compliance with the rules under industrial plants law. If the supplier and the user found a company for the purpose of chemicals management the company can become owner of the industrial plant and thus it (or its managing director) will become responsible for the compliance with the rules under industrial plants law. This applies also to obligations resulting from the plant operation permit, as this decree has an "in rem" effect, which means that the obligations are passed on from the original addressee of the decree to the new owner of the industrial plant.

An important aspect of Chemical Leasing is the closing of material cycles: The used chemicals are reprocessed and reused. In this way the quantity of wastes to be disposed is reduced. However, there is also a stage in the material cycle, in which the chemical substance is considered to be waste from the legal point of view and is thus subject to the waste management act: Namely, when the used chemical substance is returned to reprocessing. Thus, as a consequence of the implementation of the business model Chemical Leasing the provisions under waste management law become applicable to the parties involved.

In the case of Chemical Leasing with a high level of integration where the supplier takes over many services in connection with chemicals management and has therefore rented the plant of the user or if both industrial plants have been integrated in a company, thus if both plants have the same owner, the obligations regarding accompanying documents for the transport and regarding the permit for the collection and treatment of hazardous wastes are not applicable, as the Waste Management Law provides for certain exceptions for wastes generated on one's own premises.

Chapter 8

Chemical Leasing Toolkit

Petra Schwager

1. Introduction

This article presents an overview of the general content of UNIDO's Chemical Leasing toolkit, which aims at enabling a systematic approach to the implementation of Chemical Leasing business models at company level. It has been elaborated in close cooperation with the national and international experts involved in the UNIDO project "Promotion and implementation of Closing-the-Loops cooperation and business models in the chemical industry" that started in 2004. The experience obtained in Egypt, Mexico and Russia presents the basis for this instrument, which is expected to be applied by the entire UNIDO Cleaner Production network.

The toolkit covers the main steps to be undertaken to ensure smooth and efficient application of Chemical Leasing in industries of different sectors and countries. Starting from a detailed market analysis and screening process to define the most suitable sectors and companies in a given country, the toolkit provides guidelines on the implementation and monitoring of Chemical Leasing business models at plant level.

It should be noted that the present toolkit is the first attempt to provide concrete guidelines for experts and especially independent monitoring institutions like National Cleaner Production Centres (NCPCs) that are working with industry on Chemical Leasing business models. UNIDO intends to revise and update the toolkit on a regular basis and also to elaborate a specific Chemical Leasing manual in the near future.

The Chemical Leasing (ChL) approach is a new and innovative instrument to promote sustainable management of chemicals and close the material cycles between suppliers and users of chemicals ("closing the loops"). Chemical Leasing is applicable to large companies as well as small and medium size enterprises of different sectors. Experience shows that these new models reduce ineffective use and over-consumption of chemicals and help companies to enhance their economic performance.

2. Screening of potential Chemical Leasing projects (sector level)

The aim of this first step is to analyse and define chemical sectors and processes with high potential for Chemical Leasing in a given country.

- Check of industrial sectors (based on the existing supplier structures) using a set of specific selection criteria (e.g., rate of chemicals consumption within sectors, economical criteria, opportunity of a sector for growing etc.) → **short list of industrial sectors**.
- Description of the processes where chemicals are used that might qualify for Chemical Leasing projects (general technical, organisational and administrative description) → **short list of industrial processes**.
- Analysis and assessment of the economic, operational and environmental impacts related to the application of chemicals in processes:
  - Overall assessment of costs and prices of material
  - Other related costs (e.g., transportation, handling, storage, waste management etc.)
  - Technologies applied
  - Environmental impact (e.g., workplace emissions, ambient air emissions, solid waste, etc.) → **overview of economic, operational and environmental impacts**.
- Identification of main actors involved in processes within the selected sectors; develop a list of major suppliers and users and allocate them to categories (e.g., size, existing contacts, etc.) → **list of potential ChL clients**.
- Design of a database, which lists potential ChL companies, stating the respective sectors, the chemicals they use, etc.: → **ChL database** (Table 1).

Based on this database, information on processes, environmental and economic impacts and potential clients can be combined to identify processes and companies where ChL could be applied.

Figure 1 summarises three important dimensions for the screening of possible Chemical Leasing projects.

The first dimension shows the suitability of processes for the Chemical Leasing business model. A low value shows considerable obstacles for implementing Chemical Leasing (e.g., processes with a given recipe for chemicals input). A high value indicates good applicability of Chemical Leasing (e.g., knowledge sharing enables process optimisation related to unnecessary chemicals consumption).

The second dimension shows how many potential clients exist and gives an indication of potential multiplying factors. If it can be applied to many companies, a high value results.

Table 1. Chemical Leasing database

ID	Name	Function	Chemicals	Technology	Sector	Experiences

ID: Identification code
Name: Name of company
Function: Involvement (role within) in Chemical Leasing (supplier of chemicals, user of chemicals, supplier of equipment, consultant, etc);
Chemicals: Chemicals applied
Technology: Technology used in the specific process where Chemical Leasing is applied
Sector: Industrial sector of the company involved
Experiences: Space for comments on case specific experiences

a) Suitability of processes → basis for applicability of Chemical Leasing
b) Potential clients → basis for potential multiplying factor
c) Requested impacts → basis for chemical consumption, ecological and economic relevance

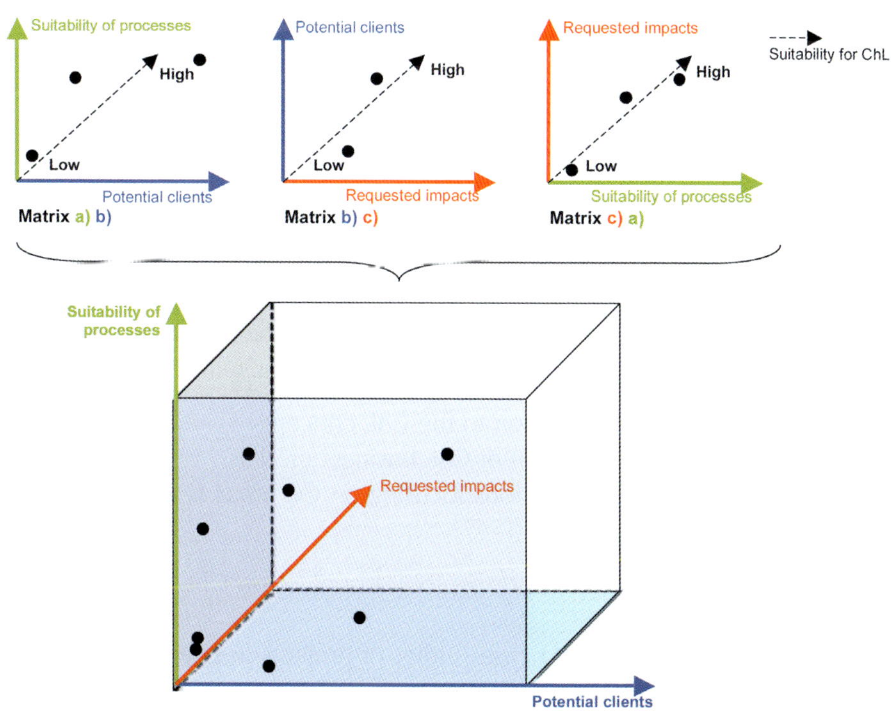

Fig. 1. Screening matrix

The third dimension highlights the impacts. Positive (economic and ecological) impacts result in high score, negative impacts such as higher risks, more hazardous chemicals, give a low score.

The position of a potential Chemical Leasing project in the matrix indicates the efforts that a National Cleaner Production Centre or an independant expert should undertake to promote the project. If all three areas show a high score, the potential for a successful ChL project is high.

3. Selection of potential Chemical Leasing clients (company level)

Three basic tools were developed to support the selection of clients that might be interested in participating in a Chemical Leasing business model.

3.1 Chemical Leasing data base

The already existing database (see structure in Table 1) should be checked using the generated list of potential clients (suppliers and users) to find appropriate partners for ChL business models.

3.2 Chemical Leasing questionnaire

As a supportive tool, it is suggested to design a specific questionnaire to investigate and get feedback on the possibility of applying the concept of ChL by companies of different industrial sectors, industry federations and chambers. This could be a very valuable source of information and a good planning tool. The general questionnaire should be structured as follows:

- Overview of and introduction to the ChL concept;
- Examples for the application of ChL business models;
- 10–15 sector specific questions focusing on the areas where ChL might be applied.

In parallel to this and as a supportive tool a booklet on ChL success stories should be developed.

Note: At present, UNIDO case studies of projects in Egypt, Mexico and Russia are already available.

3.3 Check list for Cleaner Production projects

The following check list is a support tool for experts, NCPCs and other organisations involved in Cleaner Production activities to screen existing CP projects and define the potential for starting-up Chemical Leasing business models. Legal compliance is a must for all projects.

The check list includes the following 5 steps and is designed in a way that it provides immediate reliable results.

(1) Please check which amount of chemicals are consumed by the company (based on data of the CP project)

Name of chemical	Hazardousness	Consumed amount	Points assessment for consumption
			(4) Big
			(3) Medium
			(2) Small
			(1) Very small

(2) Please assess if reduction of chemicals consumption by process optimisation can be expected?

If NO → stop the checking procedure
If YES → continue with (3)

(3) Are several users existing running the same process (potential multiplying factor) or are suppliers known that might contribute to process optimisation and that might have other clients for the same chemicals?

Point assessment	Multiplication potential	Comments
	(5) High	Several comparable users of chemicals, suppliers with broad technical know-how
	(3) Medium	Few comparable users of chemicals, suppliers with little technical know-how
	(1) Low	No comparable users, suppliers with technical know-how

(4) Please mark fields relevant for each chemical in the following assessment matrix using the results of step 1 and step 3 (Fig. 2):

	Chemicals consumption			
Potential multiplying factor	big 4	medium 3	small 2	very small 1
high 5	20	15	10	5
medium 3	12	9	6	3
low: 1	4	3	2	1

Fig. 2. Assessment matrix

(5) You can take your decision following below stated indications and based on the score obtained

3 or less points in matrix: No potential for ChL activities

4 to 9 points: Medium potential for ChL exists, take the following actions:

- Contact user and supplier and present ChL concept
- Test attitude towards ChL, if positive, undertake 1–3 visits
- Elaborate a first project idea (if necessary help to find partners)
- Discussion on potential optimisation with involved partners
- Assistance in establishing the **Chemical Leasing cooperation agreement**[1]

More than 10 points: High potential for ChL exists

- Follow actions stated above

1 This agreement is the initial paper that defines the interest of a minimum of two companies and the NCPC (and/or involved experts) to cooperate in the development of a Chemical Leasing business model.

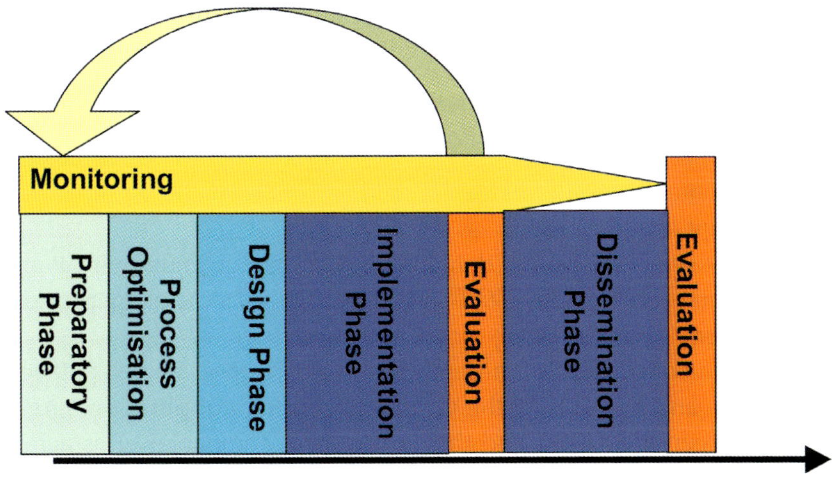

Fig. 3. Chemical Leasing project phases

4. Chemical Leasing at plant level

Successful application of Chemical Leasing business models with tangible environmental and economic benefits requires a systematic and structured approach. UNIDO defines five phases of a Chemical Leasing business model that cover a period of approximately 24 months (Fig. 3).

4.1 Preparatory phase (months 1–5)

During the first 1–5 months it is necessary for experts and NCPCs to get an overview of industrial processes and technologies applied in the company to be able to set up a detailed work plan for the implementation of ChL. Besides that, presentations to the top management and awareness raising for the staff of the company and other involved partners need to be undertaken.

Steps

- Quick scan survey of the industrial processes at plant level to get an initial overview of the processes and technologies used as well as a collection of basic data, e.g. amount of chemicals used, cost of chemicals purchased, number of pieces or other indications that might be used as *unit of payment*[2] in the Chemical Leasing contract.

2 The unit of payment is directly related to the functions performed by a chemical, e.g. metal pieces degreased, m^3 water cleaned.

- Presentation of the ChL concept to the top management of the company (might also be the first step in some cases).

  Note: The agreement of top management is essential and no further actions should be undertaken without it.

- Pre-selection and description of the processes where ChL could be applied at plant level.
- Information seminar (approx. duration: 3–4 hours maximum) on CP and ChL for involved (company) staff and partners (e.g., technology provider and supplier)
- Further data collection and in-depth analysis to support implementation
- Preparation and signature of the **Chemical Leasing cooperation agreement** between the NCPC and the participating companies. This agreement is the initial paper that defines the interest of a minimum of two companies and the NCPC to cooperate in the development of a Chemical Leasing business model.
- Training for the staff of the companies on the concept of ChL. It is recommended to include staff from the environment and quality control department, the purchasing department, the legal department (if any), technical and operational staff, and others.
- Design the work plan in detail and assign clear tasks and timeframes for completion to the responsible staff members. Staff of the training department and the financial department might also be involved at this stage.
- Upon approval, start-up of the implementation of the work plan for the process optimisation phase.

4.2 Sustainable process optimisation phase (months 6–11)

This phase follows the principles of Cleaner Production, which means the continuous application of an integrated preventive environmental strategy to processes, products and services to reduce risks to humans and the environment. It is recommended to follow UNIDO's CP toolkit[3] for the process optimisation phase, which is applied by UNIDO's Global CP Programme. The toolkit contains 10 chapters and provides guidance on how to undertake a CP assessment. It gives information about mass flow, energy flow, costs, health and safety issues, etc. This information should be used for identifying and implementing the most efficient CP options for process optimisation.

3 www.unido.org/cp

During this phase the role of an NCPC and/or an external ChL expert can vary and will depend on the agreement between the participating companies, their technical know-how and interest in getting actively involved in this phase. It might include coordination, technical support, supervision, etc.

Steps

- Technical audit and gap analysis to define initial requirements for optimisation. This should be carried out in close cooperation with the supplier.
- On the job training (OJT) for company staff (2–5 days, depending on the complexity of the processes).
- Implementation of process optimisation (e.g., process modification and adjustment) in very close cooperation with the chemicals supplier.

4.3 Design phase for the Chemical Leasing business model (months 12–16)

Following the process optimisation, the design of the Chemical Leasing business model should be initiated. Besides a detailed cost benefit analysis the essential part of this phase is the elaboration of the Chemical Leasing contract with all the related financial, legal, technical, management, implementation and monitoring issues for the implementation of ChL.

Steps

- Define the *unit of payment*, which is directly related to the functions performed by a chemical, e.g. metal pieces degreased, m^3 water cleaned.
- Carry out a detailed cost benefit analysis (CBA) to evaluate the expected environmental and economic savings of the ChL business model.
- Define a leasing rate.
- Elaborate the **Chemical Leasing contract** that includes the participating companies and the NCPC as an independent monitoring expert institution.

It is important to clearly state and define all related financial, legal, technical, management, implementation and monitoring issues for the implementation of ChL in the Chemical Leasing contract. Additionally, the responsibilities and tasks of each partner should be precisely identified. The contract should also consider the performance in terms of quality and environment.

The following elements should be included in the contract:

- 1-Contract between the two companies and the independent monitoring Institution (e.g., an NCPC or an involved ChL expert)
  - Parties of contract
  - Subject of the agreement
  - Terms of payment
  - Agreement duration
  - Signatures

- 2-Annex 1 (Terms of reference)
  - Title of contract
  - Contractors
  - General information
  - Objective of contract
  - Tasks of contactors in the Chemical Leasing business relationship
    - Tasks of Chemical Leasing service provider
    - Tasks of Chemical Leasing service receiver
  - Waste and emission management (local legislation needs to be taken into account; legal compliance has to be assured)

- Tasks of the independent monitoring institution (e.g., NCPC) to assist companies in:
  - Process optimisation
  - Maintenance
  - Monitoring
  - Provision and supervision of personnel
  - Appendix (e.g., control plan and monitoring sheets)
- 3-Annex 2 (general conditions)
  - Confidentiality of the data
  - Publication and utilisation of results
  - Use of the experiences
  - Right and place of jurisdiction
  - Right of withdrawal and contract termination
  - Signature of contract and start-up of implementation.

4.4 Implementation and monitoring phase (one year renewable)

The implementation plan should be designed by the involved experts or the NCPC in close cooperation with the participating companies and must have the full support of the top management. The following company staff members should be consulted when developing this plan:

- Top management;
- Staff of environment and quality control department;
- Staff responsible for production (including operational staff);
- Staff of material management department and supply management department;
- Staff of the finance and the purchasing departments;
- Staff of the training department.

The expert or the NCPC will be responsible for coordinating the development of the implementation plan and should therefore charge a fixed fee per year for the regular monitoring of the Chemical Leasing contracts as an independent expert institution. This will include monthly *evaluation reports* and a final external evaluation audit. The monitoring phase is essential to ensure the proper implementation of the Chemical Leasing contract and adequate benefit sharing among the involved partners. The monitoring should include the following:

- The expert or the NCPC acts as an independent institution of expertise and evaluates the established business model as well as the technical process and its potential for improvement. In this context, the expert or NCPC is monitoring the implementation of the project during a minimum of one year, following the signature of the ChL contract.
- During the contract period the companies are obliged to report to the expert or NCPC on the progress of implementation (e.g., number of coated articles per months, problems faced during the implementation, etc.) on a monthly basis. The collected data will be analysed on a regular basis and used as a basis to evaluate and eventually to improve the Chemical Leasing business model.
- The expert or NCPC will prepare unbiased monthly evaluation reports on the implementation progress, which will be based on monthly site visits and the data received from the companies. At the end of the monitoring phase, a final evaluation report will be elaborated and presented to the participating companies and partners.

4.5 Dissemination phase (starting month 17)

Encouraging other companies, sectors and countries to follow the example of successful Chemical Leasing business models is an important element to motivate the famous snowball effect. Detailed documentation of the experience obtained in the implementation and monitoring of Chemical Leasing contracts is the first step, which should be followed by a detailed analysis of the stakeholders that can support the dissemination work and strategy development.

Steps

- Detailed documentation and preparation of PR and information material
- Analysis of stakeholders to be involved in joint dissemination work:
  - chemical associations,
  - ministry of environment,
  - NGOs,
  - research institutions etc.
- Development and implementation of a national dissemination strategy, including awareness raising and information activities.

5. Conclusions

Experiences obtained so far, show that a systematic approach based on the preventive Cleaner Production idea boosts the economic and environmental success of a Chemical Leasing business model.

The involvement of an independent entity, such as an NCPC, can considerably contribute to enable adequate benefit sharing and proper monitoring of Chemical Leasing projects.

The present toolkit intends to support the work of NCPCs and independent experts to ensure the high quality of Chemical Leasing business models and the efficient use of chemicals while reducing the risks of chemicals and protecting human health.

References

Jakl T, Joas R, Nolte R, Schott R, Windsperger A (2004) Chemical Leasing: An intelligent and integrated business model with a view to sustainable development in materials management. Springer, Wien

United Nations Industrial Development Organization – Cleaner Production Programme, www.unido.org/cp

United Nations Industrial Development Organization – Chemical Leasing – homepage, www.bipro.de/_clhome/index.htm

United Nations Environment Programme – Sustainable Consumption and Production Branch, Strategies and Tools, www.unep.org

United Nations Industrial Development Organization – Chemical Leasing business models, video, www.unido.org/doc/65592

United Nations Industrial Development Organization – Cleaner Production Toolkit, www.unido.org/cp

Chapter 9

Chemical Leasing Business Models – an Innovative Approach to Manage Asymmetric Information Regarding the Properties of Chemical Substances

Cornelia Ohl and Frank Moser

1. Introduction

On 1 June 2007 the European Union regulatory framework for the Registration, Evaluation and Authorisation of Chemicals (REACH) entered into force (Official Journal of the European Union 2007). REACH aims at improving the protection of human health and the environment through a better and earlier identification of the properties of chemical substances. To reach this goal greater responsibility is given to industry to manage the risks of chemicals and to provide safety information on the substances[1]. Manufacturers, importers, as well as companies that use chemicals in volumes of one ton and more per year are required to gather information on the properties of the chemicals they apply.

In 2005, the EU chemical industry was the second leading manufacturing sector in terms of value added per employee (excluding the pharmaceutical sector). With a sales volume of EUR 436 billion it was the world's second largest chemical industry, which consisted of several multinational companies and about 26,000 small and medium enterprises (SMEs)[2]. In particular for SMEs, for which the use of chemicals is not part of their core competencies, the costs of internally conducting research and development (R&D) regarding the risks and efficient application of the chemicals is disproportionately high. The gathering of information – as required under REACH – may thus critically depend on a transfer of knowledge from the producer to the user of a chemical.

Chemicals usually are sold to the user by the producer itself or by other vendors in the supply chain[3]. Within this concept, the transfer of information

1 For further details, see: http://ec.europa.eu/environment/chemicals/reach/reach_in_brief04_09_15.pdf.

2 European Chemical Industry Council (CEFIC) web site: http://www.cefic.org/factsandfigures/index.html.

3 Hereafter referred to as the traditional sales concept.

and knowledge is consequently a crucial factor for implementing the goals of REACH. We argue that the traditional sales concept may fall short of providing incentives for relevant information and knowledge exchange[4]. If a transfer of information enhances the efficient application of the chemical and if the efficient application reduces the volume of chemicals used, this may directly lower the profits of the chemical industry. We therefore propose to supplement the traditional sales concept by Chemical Leasing business models. Such models are able to effectively manage the risk of chemical substances either by posing incentives for transferring knowledge from the producer (the lessor) to the user of a chemical (the lessee) or by assigning responsibility for the application of chemicals to the producer.

The chapter is organised as follows: In the following section 2 we focus on the traditional sales concept and the consequences of information exchange. In section 3 we introduce possible objectives of Chemical Leasing business models and in comparison to the sales concept discuss differences in information management. In section 4 we turn our attention to the applicability of Chemical Leasing business models in practice. Section 5 closes with a summary and discussion of results.

2. Selling chemicals in competitive markets

2.1 The objective of a sales contract

The traditional approach to trade chemicals in the chemicals industry is to sell the chemicals to the users. The underlying objective of the trading process is to increase utility on the buyer's side and profits on the seller's side. If the commodity under consideration shows undesired side effects on the buyer's side this is likely to affect its consumption behaviour. Regarding the potential adverse effect of chemicals on human health and the environment it is to be expected that the impact on consumption is negative. In competitive markets the price of the commodity does not react to changes in individual demand. If one or a specific group of buyers becomes aware of the negative side effects of chemicals and reduces consumption, this is therefore very likely to directly decrease the profits of the seller. In the following, we point out that these coherences may critically impact the success of information management as required by REACH.

4 In the following we use the terms information and knowledge as synonyms without maintaining that the meaning of both is the same.

2.2 Proprietorship and the allocation of risk

In trading contracts the proprietorship of the chemical, and with that the risk of chemical accidents and adverse side effects, pass over from the producer of the chemical to the user as soon as the sales contract becomes legally valid. Of course, manufacturers, and all others in the product distribution chain, can still be held liable for personal injury or property damage caused by a defective product (here: a chemical) if obligations concerning the appropriate design (design defects), manufacture (manufacturing defects), and instructions (warning defects) are neglected. However, this is not the general case. We therefore neglect liability in negligence on part of the producer as well as on part of the user. Notwithstanding this, a large number of chemicals still pose serious threats to human health and the environment. Such adverse effects include, inter alia, cancer, allergies, and the disruption of reproductive and endocrine systems (e.g., Royal Commission on Environmental Pollution 2003). At the same time, the impact of many chemicals is not fully understood yet; and the public lacks awareness of the known and unknown negative effects of released chemicals (Commission of the European Communities 2001). The initially mentioned negative impact on the consumption behaviour of fully informed buyers may thus be counterbalanced by the consumption behaviour of less informed buyers. Their consumption compared to a situation where people are fully informed is likely to be excessively.

2.3 The impact of risk on consumption behaviour – the example of DDT

An example that illustrates our conclusions regarding the consumption behaviour of less and fully informed buyers is the use of dichloro-diphenyl-trichloroethane (DDT). DDT is a broad-spectrum insecticide for malaria vector control. Although little was known about the effects of this substance to human health and the environment, the high potential of DDT to fight malaria led to a massive agricultural overuse of DDT in the last century. In 1962, the biologist Rachel Carson pointed out in her famous book Silent Spring that DDT causes cancer and harms biodiversity. This gave birth to the environmental movement in the US, which eventually led to a ban of DDT for agricultural use. In the 1970s and 1980s, people agreed that it was mainly the overuse that posed serious threats to human health and the environment (Deichmann 1972) and consequently replaced DDT in most antimalarial uses by less persistent, and more expensive, alternative insecticides. To date, DDT is regulated under the Stockholm Convention on Persistent Organic Pollutants, which permits the ongoing use of DDT under certain circumstances. This shift in preference from DDT to other pesticides with similar characteristics (Sharma 2003) is at least to some extend due to the acknowledgement of the chemicals adverse side effects and the notion that alternatives to DDT are less risky. Information management which calls producers and users of a chemical to actively gather information on the properties of their chemicals – as required by REACH – should thus be a promising route to effective risk management of chemical substances. However, below we identify one important loophole.

2.4 Information asymmetries

Small and medium enterprises (SMEs), for which the use of chemicals is not the core business, may face disproportionately high costs for gathering information on the chemicals they use. So the producer (for simplicity reasons here also assumed to be the seller of the chemical) is logically the only source the user can get information from. As pointed out above in competitive markets producers increase their profits by selling as many chemicals as possible. Consequently, if information transfer leads to a decreasing demand of chemicals, the traditional sales concept is likely to foster situations of asymmetric information. The reason is that the sales concept, as it applies here, provides no incentives for the producers to transfer their knowledge to the users. To highlight this we focus on the following setting:

The buyer of a chemical lacks in-depth knowledge on the efficient application of the chemical in use. This leads to excessive use (i.e. over-consumption) of the chemical compared to a situation where the user is fully informed. Now the producer transfers knowledge to the user, which concurrently increases the efficiency of the chemical's application. All else being equal, this lowers the user's demand of chemicals. In a competitive market this does not change the price of the chemical and thus the profit of the seller decreases. We therefore maintain that the traditional sales concept poses no incentives to fully inform the users on the properties of the traded chemicals. May this change under the REACH initiative?

2.5 A new institutional setting: REACH

With enforcement of REACH users may insist on information transfer and with reference to REACH may claim the actual information transfer by law. However, monitoring of the producer's knowledge base is usually not easily achievable. So the user may not be aware of the user's advanced knowledge on the chemical's efficient application. Moreover, REACH does not explicitly require transferring knowledge regarding the efficient handling of chemical substances and so the user may be unable to enforce such knowledge transfer by law. Accordingly, producers are offered the possibility to keep their information confidential. The incentive to keep information confidential is in particular strong if knowledge transfer lowers the consumption of the chemicals.

The example of DDT shows that decrease in demand may not necessarily be due to a transfer of knowledge regarding the chemical's efficient application. It may also happen if a third party communicates the risk of a chemical's side effect to the public. Whatever the cause of a change in consumption is, the outcome is the same: a loss of profit on the part of the seller. Accordingly, if monitoring of the knowledge base of the producer is not achievable and/or knowledge transfer not enforceable by the institutional setting, information asymmetries between the producer and the user of a chemical will persist.

Thus, the aim of REACH to effectively manage the risk of chemical substances by means of information management is not self-enforcing. Moreover, the traditional sales concept neither provides incentives for transferring knowledge to professional nor public consumers. If the costs of gathering information on the part of the buyers are prohibitive, over-consumption of chemicals is likely to remain. For a number of chemicals human health and environmental risks are positively correlated with the amount of chemicals in use (e.g., Koëter and Visser 2000). Accordingly, the traditional sales concept fails to support the aim of REACH, the enforcement of an effective chemicals risk management.

3. Chemical Leasing business models

In section 2 we have argued that for small and medium enterprises (SMEs) the costs for gathering information on the properties of the chemical substances they use are disproportionately high and for some of the SMEs even prohibitive. Increasing knowledge on the properties of the chemical substances on part of the users may consequently presuppose a willingness on part of the producers to transfer their knowledge. However, we claimed that in competitive markets the traditional sales concept fails to provide incentives for information transfer if this lowers the demand of chemicals and subsequently decreases the profits of the sellers. Moreover we pointed out that it is also not the focus of REACH to stimulate knowledge transfer with respect to the efficient application of a chemical although for a number of chemicals adverse side effects for human health and the environment are positively correlated with the amount of chemicals in use. In this section we show that *Chemical Leasing business models* are capable to deal with these shortcomings effectively.

3.1 Objectives of Chemical Leasing

Chemical Leasing business models can differ with regard to (1) *location of the application of the chemical/ownership of the chemical* – the chemical can be applied both on the site of the supplier and the user; (2) the *proprietorship of the equipment* – the equipment can be owned either by the supplier, by the user, or by a third party equipment provider; (3) the *application of the chemical* – in leasing contracts the application of the chemical can fall under the responsibility of the supplier, the user, or a third party; (4) the *operation of the equipment* – the operation of the equipment may either fall under the responsibility of the lessor or the supplier; and, (5) the *recycling and/or the final disposal* – the recycling and/or final disposal of the chemical may either be provided by the user, the supplier, or a third party[5].

Ohl and Moser (2007) have analysed two different types of leasing models which are derived on the basis of the aforementioned design options (1) to (5) – a type A that differs least and a type B that differs most from the traditional sales concept. Both models have in common that opposed to the sales concept there is no change of proprietorship; i.e., neither in model A nor B the user becomes the proprietor of the chemical as is the case when selling the chemical. The models differ as follows:

Chemical Leasing model A

As in a traditional sales contract the chemical is applied by the user (the lessee), who provides the equipment for application, applies the chemical and operates the equipment. However contrary to a sales contract and with reference to design option (5), the responsibility for the recycling and/or the final disposal is shifted from the user to the producer. This design differs least from a traditional sales contract.

Chemical Leasing model B

The chemical is applied by the producer; the equipment is owned either by the producer or by a third party equipment provider; the responsibility for

5 For details, see Jakl et al. (2004) and Schwager and Moser (2006).

applying the chemical, operating the equipment as well as the responsibility for the recycling and/or the final disposal of the chemical is shifted from the user to the producer. This design differs most from a traditional sales contract.

3.2 Selling versus leasing – a comparison under the condition of information asymmetries

The risk of negative impacts on human health and the environment increases with the amount of traded chemicals. It is therefore important to find out whether shortcomings of the traditional sales concept can be avoided by Chemical Leasing business models.

First, let us recall the shortcomings of the sales concept in face of information asymmetries. As stated in section 2, the traditional selling of chemicals, firstly, does not prevent over-consumption. In a competitive market it is rather the aim of the sales concept to sell as much commodities as possible to increase the profits of the sellers. A sales contract consequently provides no incentives to reduce such adverse effects of chemicals on human health and the environment that are related to the quantity of chemicals in use. Secondly, the sales concept provides no incentives for transferring accurate knowledge on the efficient application of the chemical, as this lowers the profits of the sellers. In contrast, Chemical Leasing business models are capable to effectively deal with problems of information asymmetries. Based on the findings of Ohl and Moser (2007), both models introduced above (A and B) show comparative advantages as follows:

Comparison of model A with the traditional sales contract

A producer offering a leasing contract in line with model A is assigned responsibility for disposal and/or recycling the chemical under consideration. This poses incentives for transferring advanced knowledge regarding the efficient application of the chemical if inefficient application increases the producer's costs for disposal and recycling. Any inefficient application of the chemical by the user likely contributes to diminishing the quantity of the solvent that can be re-used after recycling and to increase the amount of hazardous waste, which has to be disposed after the chemical's application. A leasing contract in accordance with design option five thus poses incentives on part of the producers to transfer advanced knowledge to the users on the characteristics and properties of the chemical under consideration. Moreover, if the chemical shows a potential for recycling, the producer also has a strong interest in effectively performing the recycling process as the producer (the lessor) thus can reuse the (extracted) chemical for other leasing contracts.

Comparison of model B with the traditional sales contract

In model B, the responsibility for the application of the chemicals and the operation of the equipment is shifted from the user (the lessee) to the producer (the lessor). On part of the user, this significantly reduces the probability to be held liable for damages to persons and goods. Moreover, the user is able to reduce its costs of handling the chemical to zero as the costs of handling. Such costs are usually positively correlated with the amount of used chemicals. The producer thus has an incentive to handle the chemical efficiently once this decreases its own costs of applying, disposing and recycling the chemical. Unlike model A, incentives for information exchange are not provided here. Nevertheless, efficient application, disposal and recycling are still ensured by assigning responsibility for these activities to the producer. If the producer offers a leasing contract and has advanced knowledge regarding the efficient performance of these activities it is very suggestive that the lessor will make use of this information in order to increase its own profits.

All in all, both models are able to reduce the over-consumption of chemicals due to information asymmetries between producers and users. If producers of chemical substances have advanced knowledge on the efficient application, disposal and recycling of produced chemicals the dissemination of this knowledge is neither required by REACH nor supported by the traditional sales concept. This poses additional risk to human health and the environment. These risks can be avoided by the introduction of Chemical Leasing business models, an innovative approach that delivers incentives for efficient application, disposal and recycling of chemicals.

4. Chemical Leasing business models in practice

Chemical Leasing business models show comparative advantages to the selling of chemicals. However is Chemical Leasing applicable in practice? A crucial aspect for the introduction of Chemical Leasing models is that the lessor is capable to either transfer or use advanced knowledge regarding efficient chemicals handling. Moreover, additional incentives are created if the chemicals show a potential for recycling so the producer can reuse the chemicals in further leasing contracts. If these requirements are met Chemical Leasing business models are capable of avoiding the shortcomings of and of prevailing against the traditional sales concept. In the following we highlight the importance of Chemical Leasing business models for practical application.

4.1 Relevance of information asymmetries – the example of VOCs

The focus is on a small and medium enterprise (SME) for which the use of chemicals is a minor part of the entire production chain. The SME uses painting guns for painting and coating operations of metals. These guns are cleaned with chemical substances, i.e. low volatile organic compounds (VOCs).

VOCs include organic compounds such as xylene, n-butyl acetate, or methylisobutyl ketone (National Pollution Prevention Centre 1995). The application of VOCs poses a potential risk of polluting the environment through spillages and accidents (Austrian Federal Environment Agency 2002). The most common VOCs used in painting processes are xylene, n-butyl acetate, or methylisobutyl ketone. Xylene causes various negative short-term side effects, such as chemical pneumonitis, eyes and skin irritations. The long-term side effects of a repeated or prolonged contact may cause effects on the central nervous system. If swallowed, aspiration into the lungs may result in chemical pneumonitis. Moreover, xylene is toxic to aquatic organisms (IPCS 2002).

VOCs are the only chemicals the SME applies in larger quantities so that the application of the chemicals is not part of the user's core competencies. As the cleaning process has minor importance in the overall production process of the SME, the costs for the company to conduct research and development regarding the risks and efficient application of the chemicals are disproportionately high and the producer is probably the only source of information for the user.

In contrast to this, on the part of the producer, the handling of VOCs is a core business. Advanced knowledge regarding the effective and efficient application, as well as the proper handling and secure storage of the solvent, is thus likely to be available at the producer. Neither REACH nor the traditional selling of chemicals requires the producer to transfer this knowledge to the user.

4.2 The recycling potential of VOCs

VOCs can be potentially recycled. Due to their chemical stability, non-flammability, and low latent heat of vaporisation, the method of choice for recovering VOCs from waste oils received during the cleaning operation is distillation. This recycling procedure can be easily performed on-site (The Northeast Waste Management Officials' Association 1998). The recycling procedure, however, becomes ineffective, if these waste oils are collected or handled in a suboptimal manner. First, this can be through contamination with impurities, such as paint particles and other contaminant materials, as a high concentration of impurities increases the quantity of solvent that will remain in the residue after distillation. Secondly, this can be by the blending of different types of solvents applied in the cleaning process, as resulting mixtures may form azeotropes, which are generally difficult to separate. Moreover, certain paint and solvent wastes should be carefully segregated, as otherwise the cleaning efficacy of the recycled solvent may decrease (The Northeast Waste Management Officials' Association 1998). Possible measures to minimise losses of solvent thus include:

- *The segregation of solvent and paint wastes* to avoid the formation of azeotropes to maintain the cleaning characteristics of the recycled solvent; and
- *The application of micro-filtration systems* to remove solid impurities from the solvent prior to its distillation.

It is likely that a company's low knowledge level regarding the proper handling of the chemical hampers effective recycling. Moreover lack of awareness regarding the adverse side effects of VOCs is likely to contribute to a careless handling of waste oils. Accordingly, the company may not properly segregate different types of solvent and paint wastes and consequently (if at all) perform the recycling of their chemicals ineffectively.

As with regard to REACH the transfer of knowledge regarding efficient recycling procedures is not explicitly called for, a chemical producer maximising profit has no incentive to transfer this kind of knowledge. The reason is that exchanging this kind of information would initiate the recycling of the VOCs or improve the recycling process on part of the user. Subsequently, fewer chemicals would need to be restocked by new orders of the user, thereby decreasing the producer's profits.

4.3 Prevalence of Chemical Leasing business models over traditional sales contracts

In this section we take into account that both lesser or lessee (like sellers and buyers of chemicals) are usually interested in profit maximisation. Accordingly the producer has to be willing to offer leasing, instead of a sales contract and the user has to be willing to take this offer, instead of the sales contract.

Ohl and Moser (2007) found that a producer is willing to offer leasing instead of a sales contract, if (a) the producer is able to use or to transfer advanced knowledge on the efficient application of the chemical; (b) this transfer allows decreasing its own costs of applying, disposing and/or recycling the chemical; and (c) the decrease in costs outweighs gains achieved by an overconsumption of chemicals. If these requirements are met, Ohl and Moser also showed that at the same time a price for Leasing Model A and B exists which the lessee is willing to pay and, with that, to switch from the traditional sales to a leasing contract.

A crucial issue when determining the leasing price is that as in traditional sales contracts the charge for chemicals in use (now leased instead of sold) can be specified per unit of used chemical. Transferring knowledge regarding the efficient application, ceteris paribus, lowers the aggregated charge for leased chemicals. This charge contributes to the profits of the lesser or decreasing charge of leased chemicals due to enhanced efficiency of application thus lowers the attractiveness of a leasing contract with knowledge transfer. With regard to leasing model A this loss in profit has to be outweighed by savings in recycling and/or disposal costs. If this is granted the producer is likely to switch from selling the chemical to offering leasing with knowledge transfer. Preconditions for balancing the profit gap are firstly that the producer (i.e. the lessor) has advanced knowledge on the chemical's efficient application and secondly that with the transfer of this knowledge the lessor is able to perform the recycling and/or final disposal of the chemical in a more efficient way than the user. If both requirements are met, the lessor and the lessee are able to profit more by leasing model A than by a sales contract.

Regarding leasing model B, the costs on part of the producer (in comparison to a sales contract) increase due to the assignment of responsibility for the application and safe handling of the chemical. These costs have to be covered by the price of leasing model B, which consequently is likely to exceed the price of a traditional sales contract considerably. Nevertheless this price increase is acceptable for the lessee if the cost savings resulting from shifting responsibility for applying, disposing and recycling the chemical exceed additional costs (due to the higher price) of the leasing contract. This will only hold true in cases where firstly the user applies, disposes and recycles the chemical inefficiently, and secondly where the producer (the lessor) has a head start in knowledge enabling it to apply, dispose or recycle the chemical more efficiently than the lessee. If these assumptions apply, profits gained by an increased consumption (in a traditional sales contract model) can be outweighed by a leasing price that the lessee is willing to pay.[6]

6 For details see Ohl and Moser (2007).

Regarding SMEs, improper handling of the chemicals may also cause:

- **Surplus application costs** – due to probably higher chemical consumption because of information asymmetries regarding the proper application of the chemical;
- **Surplus disposal costs** – if the solvent is applied in a suboptimal manner disposal costs will most likely increase as well;
- **Higher probability of liability claims** – lack of awareness regarding the risk of the chemical contributes to a positive correlation between the amount of chemicals in use and adverse side effects on human health and the environment. Thus the company might have to deal with more liability claims;
- **Surplus recycling costs** – a number of chemicals are potentially to be recycled. However, the improper handling of the solvent will probably reduce the output of the recycling procedure. This increases the amount of chemical waste as well as the need for further chemical production.

Chemical Leasing business models are likely to reduce such costs as well. Moreover, neither REACH nor the traditional sales concepts are in a position to change information asymmetries between users and producers regarding the efficient handling of chemicals. The reasons are that the costs of conducting R&D activities by SMEs themselves are disproportionately high and the producer cannot be forced to transfer knowledge to the users in the absence of controlling and sanctioning mechanisms. Improper handling of the chemical will thus most likely not improve. Even if REACH closes this loophole in future, Chemical Leasing business models may show comparative advantages. The reason is that even improved information exchange does not guarantee that the user's working environment provides for appropriate storage rooms for chemicals; adequate training of workers in the proper handling of the chemicals; or sufficient awareness of workers of the adverse effects of the solvent preventing them to handle the solvent carelessly.

According to Ohl and Moser (2007) this is due to the fact that SMEs may not be able to update their working environment because of disproportionately high costs for:

- Providing a safe working environment (e.g., safety equipment for workers);
- Installing and maintaining appropriate storage sites;
- Regular maintenance of machines and equipment that are used in connection with the application of the chemical;
- Training and preventive measures for workers, and
- Recycling and proper disposal of the chemical.

While it may be impossible for SMEs to change such unsatisfactory circumstances, this is likely to be different on part of the producers. The safe handling of chemicals is part of the core business of producers. As a consequence, it may be assumed that producers have comparative advantages in handling and storing the chemical and, moreover, facilities for an effective recycling of chemicals.

All in all, Chemical Leasing business models are reasonable supplements to traditional sales contracts. They can be designed in such way that they contribute to cost effective risk management by posing incentives for enhancing the efficient application of chemicals either on part of the user (Chemical Leasing business model A) or on part of the producer (Chemical Leasing business model B). With this they are capable of reducing the danger of an excess consumption of chemicals, which in a number of chemicals results in risks for human health and the environment.

5. Summary and discussion

Chemicals undisputedly play a vital role for the wellbeing of modern societies. Many industry sectors as well as private consumers strongly depend on the use and consumption of chemicals. At present, the selling and buying of chemicals therefore highly contributes to income and employment. On the other hand, chemicals pose serious threats on human health and the environment. For this reason the European Union regulatory framework for the Registration, Evaluation and Authorisation of Chemicals (REACH) was adopted and entered into force in June 2007. Under REACH, companies (users) of chemicals are required to gather information on the characteristics and properties of the substances they use. However, we argue that neither REACH nor the traditional selling and buying of chemicals provide incentives for knowledge transfers enabling chemicals to be applied more efficiently. The reason is that in competitive markets such a regime would lower the demand for chemicals and subsequently profits of producers. Against this background, a new approach is introduced – *Chemical Leasing business models*. Such models can be designed in a way that the excess consumption of chemicals is avoided. They prevail against the sales concept if the producer has advanced knowledge on the efficient application, disposal or recycling of chemicals.

REACH does not explicitly call to transfer this kind of knowledge, although for a number of chemicals, the risk of adverse side effects is positively correlated to the amount of chemicals in use. This poses a serious loophole for reaching the goals of REACH – a better risk management by means of information management. Chemical Leasing business models are capable of stimulating the efficient application of chemicals, which in turn decreases the amount of chemicals used. Such models are therefore a promising supplement to the traditional sales concept. Either by transferring knowledge from the producer to the user or by assigning responsibility for applying, disposing and recycling the chemical to the producer, these models cope with undesired information asymmetries and consequently support the goals of REACH to be met.

References

Austrian Federal Environment Agency (2002) Altlast O 55 Eumig, Upper Austria, Kirchdorf an der Krems (www.umweltbundesamt.at/umweltschutz/altlasten/altlasteninfo/altlasten3/oberoesterreich1/o55/)

Carson R (1962) Silent spring. Houghton Mifflin, Boston

Commission of the European Communities (2001) Strategy for a future chemicals policy, COM. 88 final, Brussels

Deichmann W (1972) The debate on DDT. Archives of Toxicology 29(1): 1–27

IPCS (2002) International Chemical Safety Card (ICSC) No. 0084, 0085, and 0086 (the recommendations on these cards also apply to technical xylene) prepared in the context of cooperation between the International Programme on Chemical Safety (IPCS) and the Commission of the European Communities (CEC). Date of peer review: August 2002

Jakl T, Joas R, Nolte R, Schott R, Windsperger A (2004) Chemical Leasing: An intelligent and integrated business model with a view to sustainable development in materials management. Springer, Wien

Koëter H, Visser R (2000) Work in OECD on Chemical Safety: Approaches for human risk assessment. Industrial Health 38: 109–119

National Pollution Prevention Center (1995) Case Study – Paint Spray Booth Cleaning Practices in the Automotive Industry, National Pollution Prevention Center (NCCP), University of Michigan, USA

Official Journal of the European Union (2007) Regulation (EC) No 1907/2006 of the European Parliament and of the Council. Official Journal of the European Union 50: 136

Ohl C, Moser F (2007) Chemical Leasing business models – a contribution to effectively manage the risk of chemical substances? Risk analysis. An International Journal 27(4), (forthcoming)

Perthen-Palmisano B, Jakl T (2005) Chemical Leasing – cooperative business models for sustainable chemicals management – summary of research projects commissioned by the Austrian Federal Ministry of Agriculture. Forestry, Environment and Water Management, ESPR 12(1): 49–53

Royal Commission on Environmental Pollution (2003) The long term effects of chemicals in the environment, twenty-fourth report: Chemicals in products. Safeguarding the Environment and Human Health, London

Sharma VP (2003) DDT: The fallen angel. Current Science 85(11) 1532–1537

Schwager P, Moser F (2006) The application of Chemical Leasing business models in Mexico. ESPR 13(2): 131–137

The Northeast Waste Management Officials' Association (1998) Pollution prevention in metal painting and coating operations – a manual for pollution prevention technical assistance providers, Chap 9 (www.p2pays.org/ref/01/00777/)

Chapter 10

Chemical Leasing Calculation and Profit Sharing Model

Walter Beyer

1. Initial situation

The existing traditional business models are based on the usage of tangible operating and auxiliary materials. Therefore the supplier can only achieve higher earnings by selling larger quantities or higher value-added products.

A characteristic of this model is the traditional selling/purchasing pattern involving exactly specified products at defined prices. In this context the performance of the product is only taken into account marginally.

A typical feature of this model is that on the one hand the supplier tries to sell more volumes (increasing revenue), on the other hand the user tries to control costs by buying less and/or at a cheaper price. So the basis of business is a purely economic consideration. Ecological aspects are only considered if they provide a more cost-efficient solution than the traditional one. In general, only purchasing costs are directly linked to a product, while all other expenditures are allocated to all other relevant entities as recurrent costs.

This traditional philosophy is carried on to new approaches by the new business model: Chemical Leasing. Under this new business model the goal of the supplier is to separate the service rendered from the amount of material used.

For further calculation and assessment of business cases the performance of the operating or auxiliary material and the know-how of the supplier are now decisive factors, too.

2. The innovative business model approach

For further consideration it is essential that the new business model includes the performance of a service as well as the increase in efficiency of one or more subprocesses.

Usually this is not a core competence of the user, therefore, the service offered comprises a (more or less important) process which could be integrated into the manufacturing process directly or might be an important quality factor in the production of the product.

The removal of organic coatings from metal parts, the so-called delacquering process, can be used an example. The delacquering of hooks, for example, can be done at the site of specialised companies that have optimised processes in place and the required operating and auxiliary materials available.

In several cases the delacquering cannot be sourced out and therefore it has to be integrated into the manufacturing process directly. But the users do not have the know-how about this technology and the necessary solvent mixtures.

Delacquering is a subordinate/supporting process that does not add value to the core business. Therefore budgets for optimisation and research are usually only provided reluctantly.

The new approach combines the advantages of the systems:

Know-how of the user about the production of goods and know-how of the supplier about the material, sub-processes and the recovery of the solvents.

3. Comparison of the calculation models

The traditional model is based on costs of auxiliary material (e.g. chemicals, solvents) influenced by world market prices and the competitive situation in the market. Due to the comparability of the products the user can very easily find out the most favourable offer.

Increasing the profitability of the supplier can only be realised by the expansion of the business activity since the price is fixed. It must therefore be the aim of the supplier to sell more of its product or to find new potential revenues by combining sales with services. One solution could be the new business model Chemical Leasing.

Expenditures of the user consist mainly of costs for chemicals, disposal costs, operation costs and personnel expenditures.

Subsequently the two models are outlined.

3.1 Traditional model

Total costs of auxiliary material for the user can consist of different positions. Depending on operating and auxiliary materials used and the field of application cost positions have to be adapted to the individual situation. Nevertheless, in most cases the main positions are purchasing price, disposal and transportation costs.

The following overview of possible cost positions shows that total costs can be a multiple of plain material costs.

This is the case, for example, if an auxiliary material becomes hazardous waste after application. In case of traditional delacquering the hazardous waste output of the process amounts to approximately 100% of the solvent – input and waste treatment and disposal costs are similar to the price of the new solvent-mixture.

Traditional user model

Type of cost	Currency	Amount
Material (e.g., organic solvents)		
Disposal cost		
Transportation cost (supply of solvents, transportation of waste powder)		
Salaries		
Internal storage costs		
Costs for solvents and waste management (e.g., internal costs/salaries for collecting the waste)		
Costs for quality control		
Maintenance		
Insurance		

Total amount (traditional model)

On the supplier side calculation appears to be very simple in the traditional model. There is only the sales revenue which includes all costs of production and costs of auxiliary material.

Traditional supplier model

Type of cost	Currency	Amount
Solvents cost (traditional model)		

Income (traditional model)

3.2 New business model

The calculation of the "material service" offered according to the new business model may contain the following positions. It is significant that – due to the know-how of the supplier – the necessary quantity of the auxiliary material (solvent mixture) is less than in the traditional model.

An example: The user processes an amount of solvent mixture of approximately 100 t/year for the delacquering process.

The supplier needs only 80 t for the same purpose. 60% of the solvent mixture are fresh solvents. The rest is treated by the supplier and reused in the process. The conclusion is that based on the know-how of the supplier, only 50 t of fresh solvents are needed and also 50 t are treated to recycle the remaining 30 t.

In a very simplified (incomplete) consideration the quantities and cost structures are compared.

Traditional model

Solvents	100 t	EUR	200,000
Hazardous waste	100 t	EUR	150,000
Total amount		EUR	350,000

New model

Solvents	50 t	EUR	100,000
Treatment	50 t	EUR	100,000
Waste disposal	20 t	EUR	20,000
Service, transportation e.g.		EUR	50,000
Total amount		EUR	270,000

Added Value

Cost reduction		EUR	80,000
Solvents		t	50
Reduction of hazardous waste		t	100

It is typical for the new business model that total costs and quantities needed decrease. Besides the positive economic effect the environmental effects are striking: reduction of 100% of hazardous waste while only half of the solvents mixture is required.

An overview of possible cost positions of a supplier in a Chemical Leasing contract is shown in the following. Since Chemical Leasing is not a rigid model it may be individually adapted to the actual circumstances in order to meet the targeted cost and mass reduction.

Non-recurring costs for consulting services may be added in the first phase of introduction and implementation. It can also prove useful for the start of a successful cooperation and for the promotion of mutual trust to engage an external expert to do regular checks. The external and neutral support may be helpful initially to ensure mutual economic added values.

Cost per year – new business model – supplier

Type of cost	Currency	Amount
Cost of solvents		
Cost of service (at the facility of the user)		
Costs of transportation		
Research and development		
Costs of waste recycling and treatment		
Additional costs, e.g. salaries of staff working at the facility of the user		

Total costs of ChL service "cleaning with solvents"

Simplyexpressed in summary the new business model is interesting for the user if total annual costs are less than previous traditional costs of material, disposal and transportation.

4. Sharing added values

The presentation of the calculation basis shows that the new business model can be realised successfully if total costs are reduced.

Furthermore the user is always benefitting from the knowledge of the supplier of the Chemical Leasing in the new business model.

On the other hand the economic situation of the supplier may improve. One of the reasons is that earnings from material sales are only one factor in the new business model. Factors, like the impact of the service itself, the know-how and additional rendering of service become more significant.

The actual calculation of the difference between the total costs of the two systems is subject to the rules of the market. Thus it is not possible to make general statements of how added value will be shared.

An independent consultant acting as an intermediary would support confidence building at this stage. Such a consultant would be in a position to make suggestions for a common understanding of how the added value should be shared.

Experience shows that an accompanying support of a consultant is successful and commendabel in the introduction of a new business model like Chemical Leasing.

5. Challenges of the new business model

It is an important feature of the new business model that invoicing is not carried out by kg or l of auxiliary material but that other reference measurements are necessary.

Right now it is a big challenge to find measurements which may replace traditional units of consumption of auxiliary materials.

It must be pointed out that such units cannot be defined by the two partners independently since developing a new definition would be too cumbersome and expensive.

The quest for a measurement unit correlating with the use of the auxiliary material could lead to serious problems for small users since production fluctuations can influence the consumption strongly. For this reason it will be necessary to further develop the model of Chemical Leasing.

Chapter 11

Cost – Benefit Analysis

Rudolf Schott

1. Introduction and constraints

Chemical Leasing is based on the principle of increasing the efficiency of processes using chemicals by utilising the specific knowledge of the supplier (and/or the producer) and sharing the savings between supplier and user of the chemical. The following cost – benefit analysis presents a simplified model, just taking the business relations of two partners and one product into account. Therefore the effects of Chemical Leasing to the plant(s) of the producer, the supply chain and other macro-economic effects are not included.

The theoretical examples supplied should give the reader an idea what kind of costs may be considered in case Chemical Leasing is evaluated before implementation. Main purpose is to show how "fair sharing of added-value" can be realised and what kind of factors have major impacts.

2. Cost structure in standard producer-user relationships

2.1 User costs

Costs of a chemical for the user consist of the following major factors:

1. Purchasing costs (= revenue of supplier and/or producer);
2. Quality control costs prior to usage;
3. Storage costs (storage space, handling costs);
4. Costs of empty packages and other specific waste related to the use of the chemical (handling, storage and disposal costs).

For the simplified analysis only 1) is taken into consideration. All other costs are either marginal or highly depend on the individual process of the user and should therefore not be part of general cost-benefit analyses. Nevertheless they might be important for a specific case. The possible influence of disposal costs is described in Chapter 4.4.3.

2.2 Supplier (producer) costs

Looking at the cost structure of the production of a chemical the following kinds of costs can be differentiated:

1) Variable costs: raw material and utilities (electrical energy, steam, cooling water, etc.), overhead (if calculated related to product volume, which is assumed in the example).
2) Semi-variable costs: personnel costs (production and QC).
3) fixed costs: e.g., amortisation.

For the purpose of the cost-benefit analysis semi-variable costs are considered to be fixed costs. It is assumed that the reduction of the volume of one product will not lead to a reduction of the number of operators and/or lab personnel.

The following distribution of costs might be considered to be typical (all figures in % of the price of the chemical) (Table 1).

Table 1. Cost structure producer

Costs	Share
Raw material:	50%
Utilities:	5%
Overhead:	10%
Subtotal variable costs:	65%
Personnel	20%
Depreciation	5%
Subtotal fixed costs:	25%
Total costs	90%
Gross profit	10%
Revenue:	100% (= main user cost factor)

The share of gross profit is an important figure for calculating the chances of implementing Chemical Leasing from the point of view of the producer. In general the chances of Chemical Leasing increase with a decreasing gross profit, because additional revenues by selling services (know-how) become more important.

3. Implementing Chemical Leasing

3.1 General

To improve the efficiency of a process there are two kinds of costs to be aware of:

a) Investment costs including process optimisation (consulting) and modifying of existing or new equipment.
b) Permanent operational costs to keep up the new standard and collect additional process data.

3.2 Basic data for the theoretical example and scenarios

The theoretical example is based on the following data before implementing Chemical Leasing (Table 2).

Table 2. Basic data for the example prior to the implementation of Chemical Leasing

Parameter	Costs (EUR)
Input of chemicals per year [kg]:	100,000
Price/kg [EUR]	3
Revenue of supplier (producer)	300,000
Gross profit supplier according to Table 1	30,000
Total costs supplier according to Table 1	270,000
Thereof variable costs supplier according to Table 1	195,000
Thereof fixed costs supplier according to Table 1	75,000

These assumptions lead to a revenue of EUR 300,000/year and a gross profit of EUR 30,000/year for the supplier with a cost structure of the producer as described above leading to variable costs of EUR 195,000 and fixed costs of EUR 75,000.

There are mainly 2 kinds of costs which have to be taken into account in the implementation of Chemical Leasing:

a) Investment costs (changes in equipment and/or organisation).
b) Additional permanent operational cost (for keeping up the necessary new standard, additional services provided by the supplier, controlling the output, etc.).

The following tables shows the assumptions the example is based on (Tables 3 and 4).

Table 3. Investment costs for user and supplier (producer) to implement Chemical Leasing

Investment costs	(EUR)
Producer (Analysing the process, optimisation concept):	5,000
User (investment for modifying the equipment):	20,000

Table 4. Permanent operational costs to keep up the new standard

Permanent operational costs	(EUR)
Producer (more support for user, total 10 man-days/years)	6,000
User (training of operators, total 10 man-days/year)	3,000

Further basic assumptions:

1) There is no alternative product which can be produced by the supplier.
2) The new process can be continued without changes for a total of 5 years.

The following scenarios are analysed: an increased process efficiency of 10% and 30%.

3.3 Definition of added-value and "fair sharing"

The added value of implementing Chemical Leasing is defined as:

	Theoretical savings of the user (= increasing efficiency of the process)
Minus	Amortisation and operational costs of Chemical Leasing
Plus	Difference of gross profit for the supplier before and after implementing Chemical Leasing[1]

"Fair sharing" of added-value means that the added-value is shared by the partners equally and the loss of gross profit for the producer – compared to a conventional business model – is compensated if there is any.

Both can be done by increasing the price/unit of the chemical.

4. Calculations

4.1 Limits of efficiency

Implementing Chemical Leasing only makes sense, if total investment costs and additional operating costs required to keep the new process on a higher level of efficiency are lower than the expected savings.

1 Gross profit with reduced volume minus amortisation of implementation minus operational costs of Chemical Leasing. This figure will always be negative.

a) Calculation of annual costs of Chemical Leasing

To calculate annual costs a lifetime has to be defined. A lifetime of 5 years is assumed in the following example. The following table shows the calculation of annual costs (Table 5).

Table 5. Total annual cost of Chemical Leasing

Cost factor	(EUR)
Investment costs producer according to Table 3	5,000
Investment costs user according to Table 3	20,000
Total Investment costs	25,000
Depreciation per year (assuming a lifetime of 5 years)	**5,000**
Operational costs producer per year according to Table 4	6,000
Operational costs user per year according to Table 4	3,000
Total operational costs per year	**9,000**
TOTAL COSTS OF CHEMICAL LEASING per year (depreciation + operational cost)	**14,000**

Total annual costs of EUR 14,000 for implementing Chemical Leasing equals to 4.7% of the input of chemicals in the unchanged process (EUR 300,000/year).

Annual costs of the producer can be calculated as follows (Table 6).

Table 6. Total annual costs producer for Chemical Leasing

Annual cost factor	(EUR)
Depreciation of investment (analysing the process, optimisation concept) total EUR 5,000	1,000
Operational costs per year according to Table 4	6,000
Total annual costs producer	**7,000**

Annual costs of the user can be calculated in the same way (Table 7).

Table 7. Total annual costs user for Chemical Leasing

Annual cost factor	(EUR)
Depreciation of investment (changing equipment) total EUR 20,000	4,000
Operational costs per year according to Table 4	3,000
Total annual costs user	**7,000**

In practical cases these figures may vary to a high degree and in many cases neither major investments nor high permanent costs of implementation may occur. Nevertheless it is important to keep these costs in mind to find a fair way of sharing added value.

From an impartial point of view an increase of efficiency of approximately 5% seems to define a minimum of efficiency to implement Chemical Leasing in this case. As shown later, this percentage is probably too low in reality. Generally speaking the total (implementing and annual) costs of implementing Chemical Leasing have to be significantly lower than the savings earned by the increase in efficiency.

4.2 Scenario 1: Efficiency increase of 10%

a) Producer

The producer will produce and deliver 90% of the chemicals. The producer's new cost structure will show the following numbers (Table 8).

Table 8. Costs, revenue and profit for producer after implementing Chemical Leasing with 10% efficiency increase

Cost factor	(EUR)
Variable costs (90% of EUR 195,000 according to Table 1):	175,500
Fixed costs (unchanged according to Table 1):	75,000
Total	250,500
Annual costs of Chemical Leasing according to Table 6	7,000
Total costs	**257,500**
Revenue based on EUR 3/kg and 90.000 kg/yr	270,000
Gross profit after implementing Chemical Leasing	12,500
Gross profit before implementing Chemical Leasing	30,000
Difference gross profit	**–17,500**

b) User

See Table 9

Table 9. Costs and savings for user after implementing Chemical Leasing with 10% efficiency increase

Savings and costs	(EUR)
Theoretical savings (EUR 270,000 instead of EUR 300,000)	30,000
Annual costs of Chemical Leasing according to Table 7	7,000
Theoretical net savings	**23,000**

c) Added-value

To calculate the added value of implementing Chemical Leasing the results of Table 8 and Table 9 have to be summed up (Table 10).

Table 10. Added value in EUR after implementing Chemical Leasing with 10% efficiency increase

	Added value	(EUR)
	Theoretical net savings user (see Table 9)	23,000
plus	difference gross profit producer (see Table 8)	–17,500
=	**added-value**	**5,500**

This number only represents 1.8% of the revenue of the producer or the costs of the user before implementing Chemical Leasing.

d) Fair sharing of added-value

Fair sharing means in a first step that the loss of gross-profit of the producer has to be compensated, and in a second step that the added value should be shared equally between producer and user. This results in a fair revenue of the producer (= fair costs of the user). The calculation is shown in the following Table 11.

Table 11. Calculation of a fair revenue of the producer with 10% efficiency increase

Calculation of fair revenue	EUR
Revenue based on EUR 3/kg and 90,000 kg/yr according to Table 8	270,000
Difference between gross profit before and after implementing Chemical Leasing according to Table 8	17,500
50% of added-value according to Table 10	2,750
Total = fair revenue	**290,250**

This calculation results in a new price per kg, which has to be paid by the user. This might be useful if the output changes after the implementation of Chemical Leasing (a very realistic scenario):

New price per kg (Basis 90,000 kg and fair revenue) EUR 3.225

The gross profit of the producer after fair sharing of the added value and the difference compared to the situation before implementing Chemical Leasing is calculated as follows (Table 12).

Table 12. Gross profit of producer before and after implementing Chemical Leasing with 10% efficiency increase

	EUR
Fair revenue according to Table 11	290,250
Total costs to produce 90,000 kg/yr and the implementation of Chemical Leasing according to Table 8	257,500
Gross profit after implementation of Chemical Leasing	32,750
Gross profit before implementation of Chemical Leasing according to Table 1	30,000
Gain in gross profit	2,750

Assuming the output of the process is constant – and therefore the benefit of the user is unchanged – the new costs of the user including the fair sharing of the added value are calculated as follows (Table 13).

Table 13. Net savings of user after Chemical Leasing with 10% efficiency increase

	EUR
Material costs before implementing Chemical Leasing according to Table 1	300,000
New ("fair") costs after implementing Chemical Leasing according to Table 11	290,250
Gross savings	9,750
Annual costs of implementing Chemical Leasing according to Table 7	7,000
Net savings of user	2,750

The increase in efficiency of 10% leads to an increase of the gross-profit of the producer of 9.1% and a decrease of the cost of the user of 0.9%. The ratio of 10:1 results from the basic assumption of 10% gross-profit of the producer before the implementation of Chemical Leasing.

4.3 Scenario 2: Efficiency increase of 30%

a) Producer

The producer will produce and deliver 70% of the chemicals. The producer's new cost structure will show the following numbers (Table 14).

Table 14. Costs, revenue and profit of the producer after implementing Chemical Leasing with 30% efficiency increase

Cost factor	(EUR)
Variable costs (70% of 195.000 according to Table 1):	136,500
Fixed costs (unchanged according to Table 1):	75,000
Total	211,500
Annual costs of Chemical Leasing according to Table 6	7,000
Total costs	**218,500**
Revenue based on EUR 3/kg and 70,000 kg/yr	210,000
Gross profit (loss !) after implementing Chemical Leasing	–8,500
Gross profit before implementing Chemical Leasing	30,000
Difference in gross profit	**–38,500**

b) User

See Table 15

Table 15. Costs and savings of the user after implementing Chemical Leasing with 30% efficiency increase

Savings and costs	(EUR)
Theoretical savings (EUR 210,000 instead of EUR 300,000)	90,000
Annual costs of Chemical Leasing according to Table 7	7,000
Theoretical net savings	**83,000**

c) Added-value

To calculate the added value of implementing Chemical Leasing the results of Tables 14 and 15 have to be summed up (Table 16).

Table 16. Added value after implementing Chemical Leasing with 30% efficiency increase

		(EUR)
	Theoretical net savings user (see Table 9)	83,000
plus	difference in gross profit producer (see Table 8)	–38,500
=	**added-value**	**44,500**

This number equals to 14.8% of the revenue before implementing chemical Leasing.

d) Fair sharing of added-value

The calculation of the fair sharing of added value is done like described in Chap. 4.2. The new figures according to an increase in efficiency are shown in the following Table 17.

Table 17. Calculation of fair revenue producer with 30% efficiency increase

	(EUR)
Revenue based on EUR 3/kg and 90,000 kg/yr according to Table 8	210,000
Difference in gross profit before and after implementing Chemical Leasing according to Table 8	38,500
50% of added-value according to Table 10	22,250
Total = fair revenue	270,750

New price per kg (Basis 70,000 kg and fair revenue) EUR 3.868

The gross profit of the producer after fair sharing the added value and the difference to the situation before implementing Chemical Leasing is calculated as follows (Table 18).

Table 18. Gross profit of producer before and after implementing Chemical Leasing with 30% efficiency increase

	(EUR)
Fair revenue according to Table 17	270,750
Total costs to produce 70,000 kg/yr and the implementation of Chemical Leasing according to Table 14	218,500
Gross profit after implementation of Chemical Leasing	52,250
Gross profit before implementation of Chemical Leasing according to Table 1	30,000
Gain in gross profit	22,250

Assuming the output of the process is constant – and therefore the benefit of the user is unchanged – the new costs of the user including the fair sharing of the added value are calculated as follows (Table 19).

Table 19. Net savings of user after Chemical Leasing with 30% efficiency increase

	(EUR)
Material costs before implementing Chemical Leasing according to Table 1	300,000
New ("fair") costs after implementing Chemical Leasing according to Table 17	270,750
Gross savings	29,250
Annual costs of implementing Chemical Leasing according to Table 7	7,000
Net savings of user	22,250

The increase in efficiency of 30% leads to an increase in gross-profit of the producer of 74% and a decrease in costs of the user of 7.4%. The ratio of 10:1 resultsfrom the basic assumption of 10% gross-profit of the producer before the implementation of Chemical Leasing.

4.4 Influence of major cost factors

All variations described below are based on the assumption that all other factors and relationships according to the theoretical example and all scenarios described above remain unchanged.

4.4.1 Cost structure of producer

A gross profit of of more or less than 10% as described in above scenarios leads to interesting effects:

Lower gross profits than 10% make Chemical Leasing even more attractive for the producer, because additional revenues are achieved which are not related to high variable costs (it is much likely that low gross profits of chemicals are caused by high raw material and/or utility costs).

Higher gross profits than 10% need a higher potential efficiency to become attractive for implementing Chemical Leasing from the producer's point of view. In case of gross profits of 30% a minimum efficiency rate of 20% is needed to reach approximately the same added-value like in scenario 1.

4.4.2 Implementation and operational costs of Chemical Leasing

It is obvious that the lower implementation and operational costs of Chemical Leasing are the more economic benefits are achieved by its implementation.

Due to fixed costs of production the resulting increase in efficiency of the usage process is not the same as the economic effect.

If neither implementation nor operational costs are to be considered for the implementation of Chemical Leasing, at first sight it could be expected that the increase in efficiency equals the added-value. Using the figures of the theoretical example described above it turns out that economic effects are about 1/3 lower than compared to efficiency rates – e.g. 6.5% added-value compared to 10% increase in efficiency. The reason for this phenomenon is the fixed costs of the producer lessening the increase in efficiency.

4.4.3 Other internal savings of the user

Many processes using chemicals have a high output of waste. Increasing efficiency in many cases means reducing the waste output. Typical examples are painting, degreasing and water purification processes.

The implementation of waste-handling and disposal costs into the added-value concept might be complicated in many cases, mainly because of a lack of transparency regarding the process-specific waste costs. Nevertheless if waste reduction is a major part in the increase of process efficiency, waste costs have to be taken into account and may contribute significantly to the added-value.

In the business case of painting disposal costs may reach EUR 0.5/kg and the process efficiency (coating percentage of used paint) may be around 60% of the paint input. This assumption leads to the production of 40% waste related to the input of paint[2]. Based on the input of 100,000 kg/year in the theoretical example disposal costs of EUR 20,000/yr (40,000 kg à EUR 0.5) occur. An increase of 10% process efficiency means 66% instead of 60% paint used and consequently 34,000 kg waste instead of 40,000 kg.

The resulting savings of EUR 3,000 can be considered to be significant in view of the total added-value of scenario 1 (EUR 5,500).

4.4.4 Additional profits of producer

One of the main results of implementing Chemical Leasing for the producer is having free capacities to produce alternative products instead of the one the volume of which is being reduced by the implementation of Chemical Leasing. This effect is probably more important for batch processes to produce chemicals compared to continuous processes because of the much higher efforts to change the process of the latter. But of course it is possible to compensate the reduced volume of a continuous-process chemical by new customers.

To incorporate such effects in the added-value concept a very high transparency in the producer-user relationship is required, which might be non-realistic in many cases.

Nevertheless alternative production is a major factor in the calculation and it is quantified in the scenarios under (d) "Difference in gross profit". The figure given there equals the additional profit of the producer, if the reduced volume of chemical is replaced by the production of a chemical with the same gross profit.

---

2 Only produced by paint and not considering additional waste components like filters, water, etc., which further increase the volume of waste.

5. Conclusions

The depth of a cost-benefit analysis for the implementation of Chemical Leasing depends on the mutual trust of the partners and the transparency of the cost-structures of producer and user. The method used has to be adapted to the individual case and may follow the principles presented here in a flexible manner.

Main factors are:

a) The increase of efficiency of the process in which the chemical is used. Especially the measurement of an increased output using the same (or even a higher) input is crucial.
b) Definition and transparency of investment and operational costs of Chemical Leasing – if there are any – is another major factor in defining the added-value, which is created by the implementation of Chemical Leasing. Additionally investments have to be defined regarding their lifetime in order to allow the calculation of amortisation.
c) Additional savings and/or earnings may contribute significant additional added-values and should be taken into account as precisely as possible.

Chapter 12

Business Plannning "ChemKit GmbH" – Vienna University of Economics and Business Administration

Ready for a revolution

Team: Sissi Chen, Zin Duong, Margarete Mayrhofer, Reinhild Messner, Andreas Muhr, and Eva Riebenbauer

Cooperation partner: Thomas Jakl

1. Executive summary

1.1 Business idea

Chemical Leasing is a revolutionary and new business model answering the growing ecological awareness of society and its demand on business operations to take responsibility for the results and by-products of their economic activities.

ChemKit is an interdisciplinary team of experts providing consulting to suppliers and users of chemicals with the aim of implementing a service-oriented business model for an optimised use of chemicals in industrial processes.

The service spectrum of ChemKit includes the following core services:

- Advice on technical and chemical process optimisation;
- Comprehensive legal advice on contract design (profit sharing, liability issues and much more);
- Ongoing process monitoring.

All designations of persons and functions used in the masculine form in the Business Plan shall be construed as including the female form.

Please notice that the following article of „CHEMKIT" is an abridgement of the complete business plan titled "CHEMKIT – Ready for a revolution". For the full version and further details please contact the authors.

This concept offers clear benefits for all parties involved:

- Chemical suppliers: generating profits from selling their know-how, participation in the proceeds from Chemical Leasing co-operations after implementation, improved customer loyalty, and many more.
- Chemical users: cost reductions, efficient use of resources, reduced administrative burden, and many more.

Just as a master builder sells the hole in the wall to a customer rather than merely a sledgehammer, the focus of Chemical Leasing is on selling the full service package of chemicals usage to the customer of the chemicals manufacturer – and not just the chemicals themselves.

With every completed project, **ChemKit** generates EUR 8,000 (SMEs) or EUR 11,600 (large-scale projects) in revenue in the year of implementation.

1.2 Market and competition

ChemKit is the only consulting company in the environmental field in the Austrian market specialising exclusively in Chemical Leasing and offering its customers a tailored service package in close co-operation with the chemical supplier.

Surveys and the performance of analogous markets suggest that approx. 50% of Austrian companies for which Chemical Leasing is a viable option will have adopted this model within 10 years. This equates to a market volume of approx. 2,000 projects.

An estimate of the total market volume and the total market share of **ChemKit** shows the following market share distribution in the first years (Fig. 1).

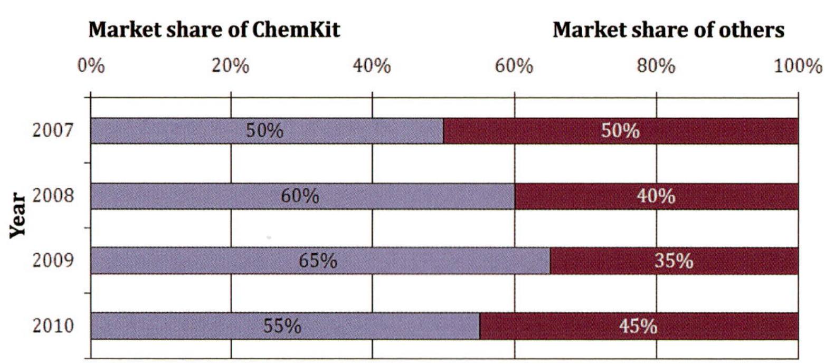

Fig. 1. Expected development of market shares in the Austrian target market (likely case)

1.3 Opportunities and risks

The macro-trend towards sustainable environmental protection and social responsibility, also on a legislative level, offers **ChemKit** good prospects for a steady business success. However, in the first years the business model depends strongly on subsidies from the Austrian Federal Ministry of Agriculture, Forestry, Environment and Water Management (BMLFUW). Any unforeseen short-term loss of these subsidies would have an extremely detrimental effect on the revenue situation of **ChemKit**.

1.4 Marketing and market entry

The measures for actively raising awareness at the market entry stage include, among others, advertisements in special-interest magazines, presentations at trade fairs, training programmes, and the participation in information events. The BMLFUW's subsidy schemes also add to the incentives for potential customers to adopt the business model.

1.5 Finances

Assuming total revenues of EUR 235,000 in the first year of operation, revenue is likely to increase to EUR 2 mio in subsequent years until 2010 (Fig. 2).

The fact that the break-even point will be reached in October 2007 exemplifies the potential success of this innovative business model (Fig. 3).

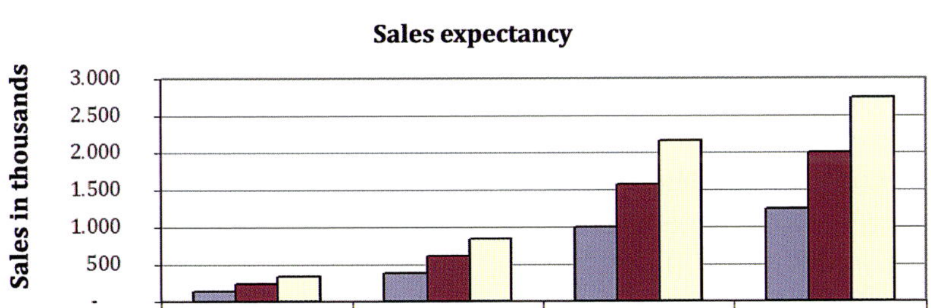

	2007	2008	2009	2010
Worst Case	143	388	1.004	1.253
Likely Case	235	608	1.584	2.007
Best Case	340	836	2.167	2.751

Fig. 2. Expected ChemKit revenue progression 2007–2010

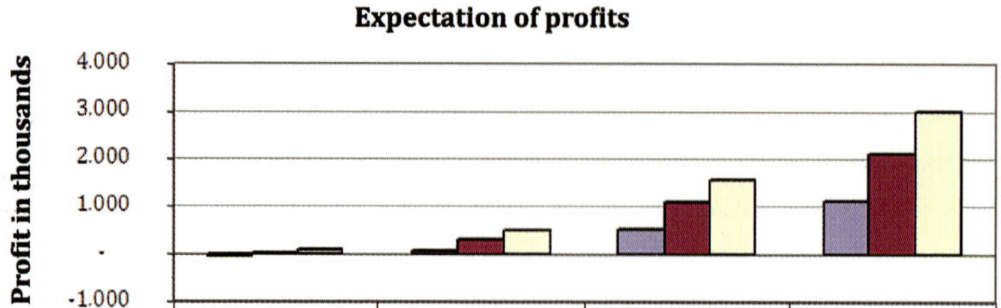

	2007	2008	2009	2010
Worst Case	-52	78	526	1.132
Likely Case	27	301	1.096	2.142
Best Case	98	497	1.571	3.018

Fig. 3. Expected ChemKit profit progression 2007–2010

Best case	Likely case	Worst case
EUR 51,000	EUR 51,000	EUR 76,000

Fig. 4. Financial requirements in the start-up year

The initial profit of EUR 27,000 will be followed up by a further positive development of profits. Revenues will increase six-fold in the third year, and exceed EUR 2,000,000 in the fourth year.

Depending on the underlying scenario, the financial requirements for the start-up year (2007) will be (Fig. 4).

1.6 Further strategy

In order for ChemKit to maintain its pioneering position in the environmental consulting market and to ensure long-term success, the company must push ahead to become established quickly in the Austrian SME market. The BML-FUW and the Austrian Federal Economic Chamber (WKO) are key cooperation partners in this regard. The calculated ChemKit growth perspectives take into account an extension of the group of target customers to include large-scale companies and an expansion into the European market and also, at a later stage, into the global market, as well as intensive co-operation with international organisations (especially UNIDO) and the enlargement of the company's own team of consultants. Chemical Leasing is the future – and the competition is not sleeping!

2. The company

2.1 Idea

ChemKit has discovered a need among many companies for professional and expert advice for the complex and expensive adoption of Chemical Leasing schemes. Based on the business model introduced above, ChemKit, as an innovative environmental consultant, aspires to bring together chemical suppliers and chemical users for the sole purpose of implementing Chemical Leasing schemes.

According to a study conducted by the BMLFUW, the implementation of the Chemical Leasing concept would be suitable for approx. 4,000 Austrian companies (mostly SMEs).[1] *A specially conducted market analysis*[2] *showed that, in the start-up stage, the SME segment would be the main target area for the Chemical Leasing consulting business of ChemKit. Large-scale companies are either not yet showing much interest in implementing the scheme or are already optimising their business processes in a way that is similar to Chemical Leasing.*

ChemKit has set itself the aim of raising awareness among the broad spectrum of Austrian small- and medium-sized enterprises with regard to the advantages of a more efficient use of chemicals and to provide information in order to allay any reservations there may be regarding this scheme, in order to secure its pioneering role in the Austrian environmental consulting market.

2.2 Company profile

Company profile: ChemKit GmbH, *legal form:* limited liability company acc. to Austrian law, *location:* Vienna, *founded:* January 2007, *main business activity:* consulting for Chemical Leasing schemes

The guiding principle of the founding team is reflected in the company motto "ChemKit – Ready for a revolution": bringing together users and suppliers of chemicals with a revolutionary, service-oriented business model.

The highly motivated and superbly qualified interdisciplinary founding team provides its customers with innovative and revenue-optimised solutions. Each team member has extensive knowledge of business, chemistry and

1 Dr. Thomas Jakl, Austrian Federal Ministry of Agriculture, Forestry, Environment and Water Management (BMLFUW), 2005.

2 Internet survey of 49 Austrian large-scale companies (for details, see Appendix).

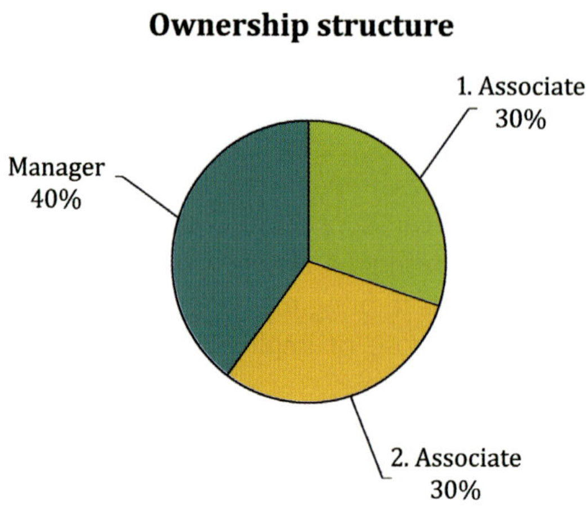

Fig. 7. Planned business shares of ChemKit

chemical engineering and the team is furthermore supported by external specialists (legal and tax advisors).

Concerning the business shares of the founding members in the initial stages, we consider the following distribution appropriate (Fig. 7).

2.3 Location analysis

We consider a central business location in Vienna to be ideal for a consultancy such as ChemKit. Apart from good travel connections, another point to be taken into account is sufficient space for future expansion opportunities with a special focus on the transition countries in Eastern and Central Europe. Moreover, many public institutions such as the BMLFUW and other partners are also based in the capital. Further locations in the various Austrian federal provinces should also be considered for the planned expansion of business activities over the subsequent years. Priority should be given to Styria, Lower Austria and Upper Austria, as key customers are located in these federal provinces.

2.4 Business objectives

The innovative concept which aims at enhancing resource productivity by changing from a mere sale of chemicals to a sale of comprehensive service packages, coupled with the high future relevance of a discerning, sustainable

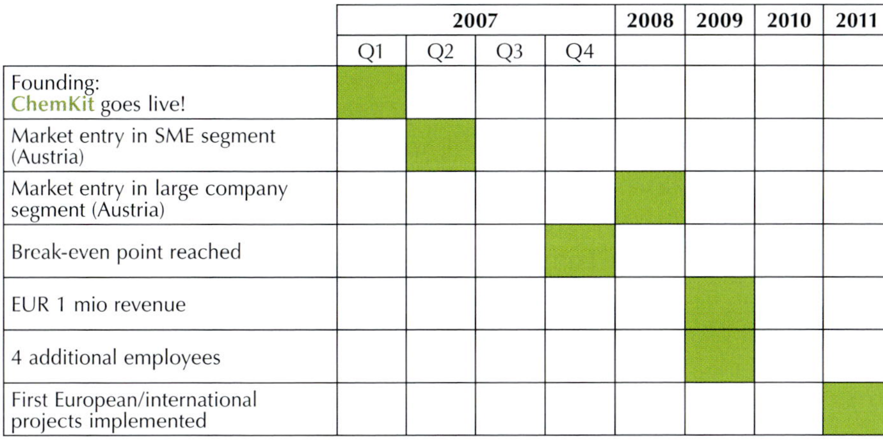

Fig. 8. Short-term and medium-term business objectives

and efficient handling of chemicals, provide the foundation for future growth considerations of **ChemKit** (Fig. 8).

Once **ChemKit** has become established as a consultancy in the domestic SME market, the next step will be to launch consulting co-operation projects with large companies as the awareness for **ChemKit** and its business model grows. Interdisciplinary expert know-how, comprehensive project experience and precise knowledge of the Austrian Chemical Leasing market are the crucial pre-requisites and convincing sales arguments here. Co-operation with the BMLFUW and the Austrian Federal Economic Chamber (WKO) is essential to boost business acceptance in the Austrian market. In addition, raising public awareness of the efficiency and efficacy of Chemical Leasing by engaging in intensive PR work will serve as an important factor of success.

The fact that Chemical Leasing has already been implemented successfully in several pilot projects[3] around the globe strengthens the **ChemKit** team in its vision to extend the business model and include the European market and later the worldwide market. The primary emphasis of these expansion models is on projects in Eastern Europe and co-operation in transition countries and developing countries. These growth efforts will, on the one hand, rely on the increased readiness to innovate and invest in young and aspiring economic systems (e.g. India, China, Russia, and many more), as well as on the opinion of experts, who see the biggest potential for Chemical Leasing in regions where the chemicals industry is only now starting to establish itself.

3 Projects by UNIDO (United Nations Industrial Development Organisation): Mexico, Egypt, Russia (for details, see Appendix).

Main success factors of ChemKit
• ChemKit consultants are highly qualified and have great market knowledge
• Synergies through networks (BMLFUW, WKO, UNIDO, and many more)
• Business acceptance and interest in Chemical Leasing
• Intensive PR work to raise awareness
• Access to subsidies (BMLFUW, WKO)
• Innovations and a head start in know-how over the competition

Fig. 9. Summary of the relevant success factors for achievement of objectives

It is still possible to shape these industries and entrepreneurs are still much more inclined to implement new, unconventional business ideas. At this stage of its expansion, ChemKit is especially interested in co-operating intensively with international organisations such as UNIDO. In its function as a "standby advisor", the team is always ready for Chemical Leasing consulting missions in transition countries and plans to use this approach to become established in international markets (Fig. 9).

3. Services

3.1 The range of services

ChemKit offers the chemical user competent and comprehensive support when opting for Chemical Leasing. An optimal service package is tailored to the needs of each customer through close co-operation with the chemical supplier, to ensure a smooth and rapid implementation of the new business model (Fig. 10).

The **ChemKit** service package comprises:

- Comprehensive information,
- Technical and chemical process optimisation,
- Advice on, or implementation of the efficient use of chemicals,
- Comprehensive legal contract design:
  - Clarification of liability issues
  - Fair profit-sharing arrangements
  - Protection of intellectual property
  - Waste law
- Consulting and implementation of information management systems,

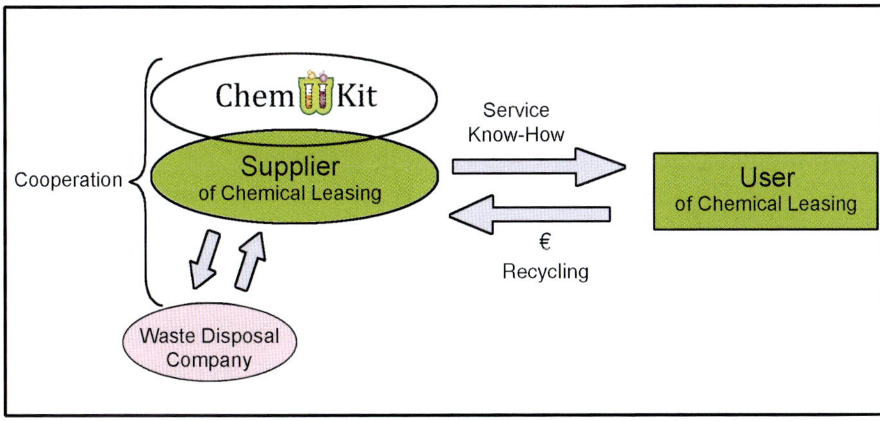

Fig. 10. The new Chemical Leasing business model of ChemKit

- Subsidy consulting,
- Employee training,
- Implementation and execution of continuous process monitoring.

3.2 Target customers and their needs

In a study conducted by the BMLFUW, 4,000 Austrian companies (mostly SMEs) that use chemicals in their operations and for whom Chemical Leasing would be a viable option were identified.[4] **ChemKit** will focus on these SMEs in the initial stages in order to implement the new business model in Austria. The advantages of this focus are that SMEs, having relatively little internal resources, depend more strongly on the support of external partners. The great number of small- to medium-sized projects will provide **ChemKit** with a wealth of experience that will be useful for future expansion and large-scale projects. Moreover, **ChemKit** will thus be able to established itself in the market and build a good reputation within the industry.

The possible applications of Chemical Leasing are extremely heterogeneous and diverse; furthermore, there are no hard and fast rules as to the optimal design of a process. Therefore, the changeover to Chemical Leasing requires information about the new business model, information on how to bring together suitable contract partners, chemical and technical consulting for process optimisation, the highest possible logistical efficiency and detailed, cus-

4 Jakl et al. (2003).

tomised contract design. The great number of applications and the necessity for customised process and contract design show that only a team of consultants with high competence in all chemical and technical matters as well as knowledgeable of business and legal issues can meet these requirements. And **ChemKit** offers precisely these services to its customers.

According to a survey of the BMLFUW and interviews with experts, the reservations that prevent users most from adopting the system is a lack of information on the existence and application models of Chemical Leasing, a mistrust of "external consultants", and, in some cases, the fear among a user's employees of losing their jobs. **ChemKit** will remedy the information deficit through targeted marketing activities. The mistrust among users and their employees with regard to consultants who have only a business education as credential is balanced out by the chemical and technical competence of the **ChemKit** team of consultants. While the employees' fear of losing their jobs upon implementation of a Chemical Leasing scheme may be justified in a small number of cases, this decision is solely up to the user. Fundamentally, the new model of chemical usage is merely intended to make things easier for the user and its employees: time-consuming secondary tasks that are inefficient because of a lack of know-how can be outsourced and the company can focus on its core competencies.

3.3 Customer benefit

A customer implementing Chemical Leasing can expect to save an average of EUR 6,000 a year in business, HR, capital, materials management and transaction costs in addition to saving consequential costs of non-quality.[5] Although the use of chemicals in the customer's company may not be part of its core competencies, it does constitute an indispensable and necessary contribution to production in the company's business. By changing over to Chemical Leasing, the user can focus entirely on its core competencies by outsourcing the chemical processes, while at the same time eliminating costly, environmentally harmful waste of chemicals.

Changing over to Chemical Leasing bears a further special benefit: The user is no longer bound by obligations under the strict directives of the future European chemicals policy, REACH (Fig. 11).

5 Jakl et al. (2003).

Customer benefits
✓ Cost reductions
✓ Environmental protection
✓ REACH compliance
✓ Focus on core competencies
✓ Process optimisation
✓ Availability of chemical and technical know-how
✓ Enhanced safety at work

Fig. 11. Overview of customer benefits of Chemical Leasing (source: Jakl, 2003)

3.4 The role of the chemical supplier

In order to implement Chemical Leasing in the user's company, ChemKit aims to establish a type of co-operation with the chemical supplier (the manufacturer, trader or mixer of chemicals) in which the chemical supplier provides the chemistry know-how and ChemKit provides the process engineering know-how, among other things. By pooling both competencies, the user is provided with an economically and ecologically attractive service package, which is furthermore customised to the user's individual needs. This allows for a rapid and smooth changeover to Chemical Leasing.

The user is not the only one to profit from this close co-operation. The chemical supplier and **ChemKit**, too, profit from this model, as will be shown below.

The chemical supplier does not incur any financial losses from the reduced quantity of chemicals sold. Now the company sells the effect of the chemicals. By making its specific know-how available – in practice this is often an unpaid service – the supplier refines its product, for which it is now suitably remunerated. The chemical supplier gets a share of the profit and the lower the volume of chemicals used, the higher the profit it derives from the project. Its payment is based on "number of pieces chemically treated, cleaned m^2", etc.

The new directives of the European chemicals policy (REACH) constitute a great challenge for chemical suppliers, because all chemical substances produced, used or imported by the company now have to be registered, evaluated and authorised. However, according to industry experts, it is especially the SMEs that do not have suitable (knowledge) resources to be able to meet these requirements without external support. This is why **ChemKit** supports its co-operation partners with in-depth REACH consulting.

Benefits for the chemical supplier
✓ Sale of its hitherto unpaid know-how
✓ Higher profit if smaller volumes of chemicals are used
✓ Environmental protection
✓ REACH compliance
✓ Long-term business relations
✓ Better identification of customer needs
✓ Precise planning of the company's own consumption
✓ Reduced material costs through lucrative recycling

Fig. 12. Overview of the benefits of Chemical Leasing for chemical suppliers (source: Jakl, 2003)

In addition, the chemical supplier benefits from the close co-operation with the chemical user, which brings about long-term business relations characterised by mutual trust. The supplier is in a position to adjust optimally to the specific needs and problems of the user to plan its use of chemicals in more detail and to discover new developments or changes in the customer's needs at an early stage.

The used chemicals are taken back by the chemical supplier for treatment, provided it has a waste treatment facility. Such a facility could also be located on the user's premises; this differs from case to case. If no waste treatment facility exists, **ChemKit** will put the supplier in touch with a suitable waste disposal company with whom it can work together. Parts of the waste output are thus recycled and fed back into the user process, closing the loop of the Chemical Leasing model. Since the chemical supplier will need to use fewer new chemical substances in the future, it profits from the cut in material costs. In addition, it makes a sustainable contribution to environmental protection (Fig. 12).

The Chemical Leasing partners can be brought together in two ways. Either **ChemKit** uses its existing networks in the industry to win the best-suited chemical supplier as a co-operating partner for the user, or the chemical supplier acquires the customers itself. In the latter case, **ChemKit** profits from its existing customer relations. By recommending **ChemKit** to its customers, which brings about a multiplier effect, it provides a rapidly growing circle of additional customers for **ChemKit** and supports the build-up and development of networks.

Growth drivers in the environmental consulting industry
★ Trends: Environmental protection by way of environmentally-oriented business management, sustainability, corporate social responsibility
★ Resource and energy price hikes
★ Sustained high demand for the optimisation of production processes and resource-efficient use of raw materials (cost reduction)
★ Economic growth
★ New markets through increasing internationalisation (Eastern Europe as a growth market)
★ New EU directives and therefore increased need for consultants (REACH)
★ Public subsidies

Fig. 13. Overview of the growth factors in the environmental consulting industry (sources: Martinuzzi (Martinuzzi André: Erfolg im Öko-Consulting, September 2005), FEACO (European Federation of Management Consultancies Associations. Survey of the European Management Consultancy Market, 2004), BDU (Bundesverband Deutscher Unternehmensberater. Facts & Figures zum Beratermarkt 2005/2006)

4. Industry & market

4.1 Industry analysis

ChemKit, which offers environmental, technical consulting, is part of the environmental consulting industry. Environmental consulting providers include technical consultancies (e.g., civil engineers, technical consultancies, laboratories), management consultancies (e.g., management consultants, data processors, tax advisors, PR consultants, market researchers) and interdisciplinary environmental consultants.

One source provides promising numbers: For his book "Erfolg im Öko-Consulting" (Successful environmental consulting), André Martinuzzi had a close look at the industry on the basis of four surveys from 1991 to 2001. He documented approx. 400 Austrian companies that offer environmental consulting services and generate a total annual revenue of EUR 276 mio. Throughout the study period, environmental consulting revenue grew by 24.5%.[6]

Due to the growing importance of a sustainable and responsible use of raw materials (i.e., also chemicals), and the effort to optimise business and production processes, as well as to minimise the waste of resources, there is an increasing demand for consultants that are both environmentally and economically oriented (Fig. 13).

6 Martinuzzi André: Erfolg im Öko-Consulting, September 2005.

4.2 Market analysis

99.6% of the Austrian market comprises small- and medium-sized enterprises, or some 300,000 companies in absolute terms.[7] Of these, approx. 4,000 small- and medium-sized enterprises have been identified for which the new business model would be suitable, taking into account their chemical applications.[8] These companies are an attractive target customer group for **ChemKit**, as they depend on external consulting due to lacking internal (financial and personnel) resources. Dr. Reinhard Joas (managing director of BiPRO GmbH) also emphasised in his interview, which was conducted in Munich, that SMEs are dependent on supporting intermediaries, and therefore constitute a great springboard for the sustainable introduction of Chemical Leasing in the market. For **ChemKit**, this focus on SMEs is ideal, as it is able to gather experience with small-scale projects and build a higher profile in the industry. In order to be able to estimate the companies' interest in the implementation, **ChemKit** carried out a poll based on snowball sampling (see Appendix). Approx. 60% of all respondents in this survey expressed their interest. The main arguments for changeover were possible cost reductions, outsourcing of processes subject to regulations and the benefit for the environment. None of the respondents had heard of Chemical Leasing before, which reflects the lack of dissemination of information in the Austrian market (cf. Jakl et al., 2003).

A comparison of market penetration with the similar Responsible Care Model shows that more than 50% (in terms of value) of the chemical companies had implemented this new model after a period of 13 years. The Responsible Care Model, under which the chemical supplier ensures that "its" chemical is used in a way that is as environmentally friendly and protective of health as possible, is based on a similar idea and is comparable in nature. However, Responsible Care by no means requires such a close co-operation between the chemical supplier and the chemical user, and unlike Chemical Leasing, does not aim at minimising the volume of chemical substances used. By analogy, it can be expected that Chemical Leasing will have been adopted by half of the target customers, or approx. 2,000 SMEs, within roughly 10 years.[9] Based on this, a diffusion curve (see appendix) was drawn up, which results in the following project demand for the first four years (Fig. 15.)

Before **ChemKit's** entry into the market, several innovators had already carried out Chemical Leasing pilot projects, and according to experts, the number of companies that have implemented the new model without an-

7 SME Forschung Österreich 2005, http://www.kmuforschung.ac.at/index.asp?j=de/start.htm.

8 Ecotec study.

9 FCIO, sustainability report on the chemical industry, 2005.

Risks within the eco-consulting industry
★ High market dynamics
★ Threat of new market entrants due to low entry barriers

Fig. 14. Survey of risks within the eco-consulting industry

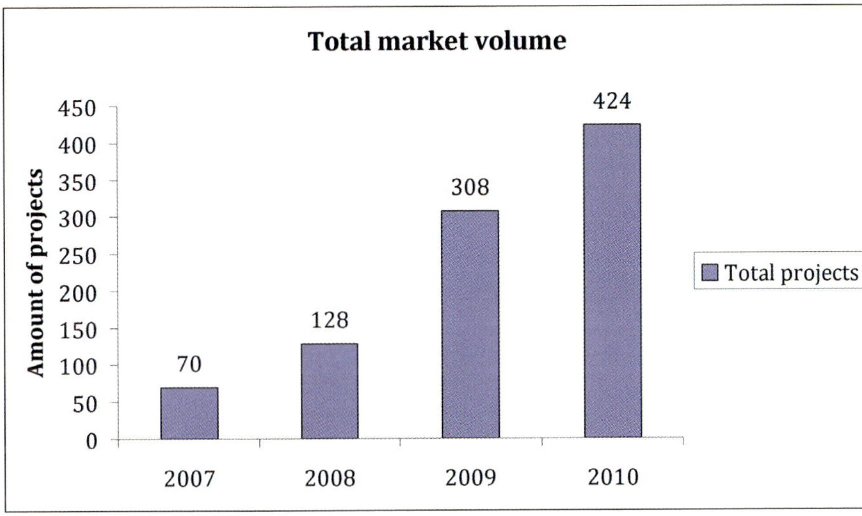

Fig. 15. Expected total market volume 2007–2010 (number of projects)

nouncing this must also be taken into account. The demand for the 70 projects in the first year of business will therefore come from "early adopters". In the following years, the number of projects will rise continuously until maturity is reached in 2010. Subsequently, demand will diminish gradually due to the beginning market satiation.

ChemKit will initially be the only consulting company in the Austrian market to be exclusively specialised in Chemical Leasing. Since there are only three competitors in the market who offer this service, the market share in the 1st year is estimated to be 50%. In the following two years, ChemKit will make further gains in market share because of its growing reputation and expert knowledge. However, the competition does not sleep, and so new competitors will enter the market. Therefore we assume that the market share of ChemKit will begin to decline as of the 4th year (Fig. 16).

The following diagram shows the distribution of projects to ChemKit and competitors for a period of 4 years (Fig. 17).

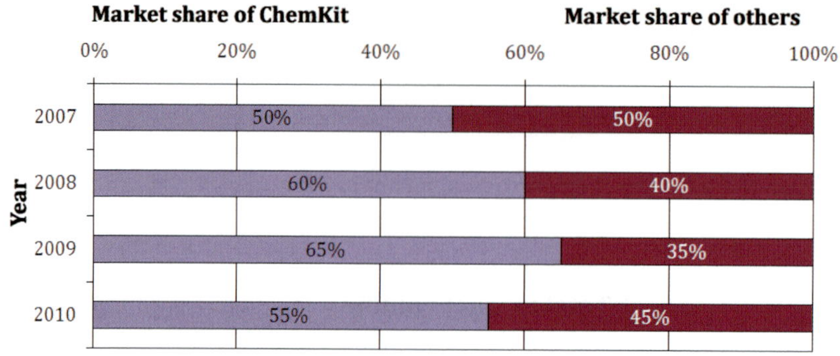

Fig. 16. Expected development of market share 2007–2010 (percent)

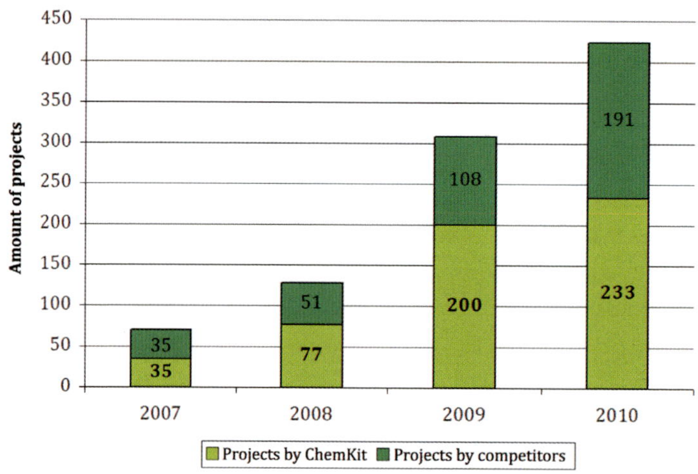

Fig. 17. Expected development of market share 2007–2010 (number of projects)

The market can be represented according to chemical application. This classification results in a total of twelve segments, each with a chemical savings ratio ranging from 5% to 35%. Owing to its comprehensive expert knowledge, ChemKit is in a position to provide consulting to companies of all the segments, and will therefore not focus on a single area. Naturally, ChemKit will specifically target the processing of those market segments with the highest savings potential (etching and casting) (Fig. 18).

By changing over to Chemical Leasing, users have average potential savings of EUR 6,000 per year. This figure does not include the possible savings in volume of chemical substances used, but concerns the costs of material flow management, logistics, transaction costs and waste removal costs.

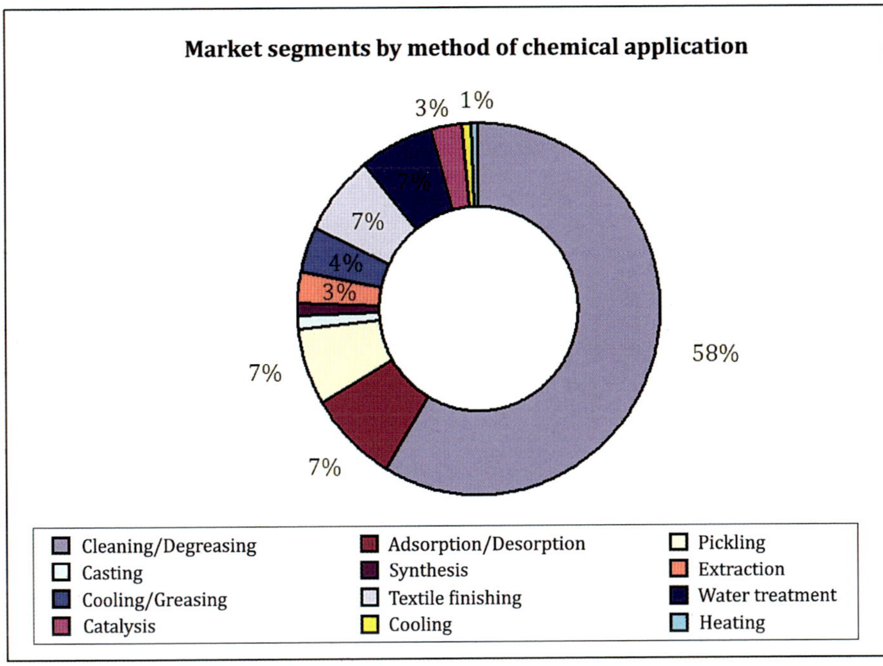

Fig. 18. Market segments according to application

4.3 Competition

Below is a list of companies offering consulting services in the Chemical Leasing sector (Fig. 19).

As can be seen from the table above, the competitors in the Chemical Leasing sector have already gained their first experience with pilot projects, they are present in the market and all enjoy a good reputation.

Therefore, the biggest challenge for ChemKit in the initial stages will be to win suitable reference customers in order to become established in the market. However, as soon as ChemKit has secured a foothold, there is nothing in the way of rapid customer acquisition, as only four consultancies service approximately 4,000 potential customers in the Austrian market.

ChemKit is the only consultancy which has accumulated the specific technical know-how in the sector, as well as all the experience from studies of pilot projects and focuses 100% on the implementation of Chemical Leasing projects. Due to its concentrated knowledge, specialisation and competence, ChemKit is highly differentiated from its competition.

Competitors	Location(s)	Service	Target customers	Market position	Strengths	Weaknesses
ChemKit GmbH	Vienna	Consulting and problem-solving with a specialisation in Chemical Leasing	Austrian SMEs in the segment of cleaning/degreasing	Growing, young enterprises	Collaboration with UNIDO and Ministry of Life 100% focus on Chemical Leasing knowledge pool and know-how accumulation	Lacking market establishment, no reference customers
BiPRO GmbH	Munich	Support and consulting in the fields of technology, business, the environment, and health	Ministries and authorities, as well as companies and federations	International market coverage	Collaboration with UNIDO, EU and BML-FUW Co-founder of the Chemical Leasing model International experience with Chemical Leasing Good market reputation	Has only completed one project in Austria so far
Beyer Umwelt + Kommunikation	Vienna	Business and communication consultancy specialising in environmental protection with specialisation in: management systems, waste management, subsidisation, Chemical Leasing and communication	Ministries and authorities, as well as companies and federations	Established Austrian chemical consultancy with a good market reputation	Experience in Chemical Leasing (Tiefenbacher GmbH) Experience in contract design for micro enterprises Constantinus Award 2005	Experience so far limited to five pilot projects
Denkstatt Umweltberatungs- und Management GmbH	Vienna Budapest	Consultant for services in the sustainable development and environmental management field	Medium and larger companies in the automotive, paper, chemicals, mineral oil, foodstuffs, electronics, steel, metal-processing, wood-processing and building materials industries, energy generation and service companies	One of the largest consulting companies in Austria	Interdisciplinary team Experience in the Chemical Leasing field (pilot project Borealis AG) Growth (one office in Budapest)	

Fig. 19. Overview and comparison of ChemKit competitors

As soon as Chemical Leasing has become a standard business model and is met by growing demand, more competitors will inevitably enter the market. By that time, ChemKit will already have set up extensive networks and enjoy an excellent reputation.

5. Marketing

5.1 Market entry

Presence in special-interest media

The founding of ChemKit is accompanied by information events about Chemical Leasing. Awareness of this business model is actively raised and accompanied by presentations in special-interest publications and representations at trade fairs. Workshops, conferences and presentations for the relevant sectors will round out the market entry in this regard. The main aim is to clearly structure cost and image benefits (sparing of resources, environmental benefits, sustainability and responsibility) for all parties and to portray ChemKit as a specialised consultancy that provides successful implementation. ChemKit and Chemical Leasing should become synonyms in people's minds.

The target group consists of all relevant sectors and companies in whom Chemical Leasing can be applied. Therefore, the focus will be put on the twelve segments identified so far (cleaning/degreasing, etching, lubricating, etc.). Relevant industries would be, for instance, the metalworking industry, the automotive industry, the paper, furniture, textiles, and plastics industry, etc.

One example for trade fair presentations would be the "Vienna Tec[10]" international trade fair for industry in Vienna (held for the first time in October 2006) or "ACHEMA[11]", the international exhibition conference for Chemical Technology, Environmental Protection and Biotechnology in Frankfurt (which takes place every three years and is the largest trade fair for chemical facility construction. The organiser of this event, "DECHEMA" also hosts smaller subsidiary trade fairs in Beijing, America and Australia; the next event is scheduled for May 2009).

Furthermore, ecological trade fairs offer the ideal stage for presenting the Chemical Leasing model and ChemKit as the competent partner for implementation. Here, the existing interest of trade fair visitors in environmentally

10 www.vienna-tec.at
11 www.achema.de

friendly solutions for production companies is used to offer them not only an environmentally friendly, but also a lucrative model for the use of chemical substances.

Co-operation with the Austrian Federal Economic Chamber, the Austrian Federal Ministry of Agriculture, Forestry, Environment and Water Management (BMLFUW) and UNIDO

The existing network of **ChemKit** business contacts and the individual partners' contacts with industry experts is seen as another pillar of **ChemKit's** market entry. These boost both the credibility and the reputation of **ChemKit**. Key co-operation partners such as the Austrian Federal Economic Chamber are important as door-openers to potential **ChemKit** customers. Owing to their experience with introducing new ideas and models and their large database, they help **ChemKit** to quickly and efficiently contact its relevant target audience.

One important public co-operation partner is the Austrian Federal Ministry of Agriculture, Forestry, Environment and Water Management (BMLFUW), which has been interested in a country-wide implementation since the model was first devised. The pilot projects carried out in Austria to date have often been initiated by the BMLFUW and constitute a stock of relevant reference project examples of their home country for the potential customer base. Until now, all companies have recorded significant savings.

UNIDO is also seen as one of the most important partners of **ChemKit**, as it already has a wide spectrum of international experience in Chemical Leasing. Apart from an exchange of experience, expectations are that international contacts will be established, which are essential for continued growth.

Internet presence

As one of the main sources of information, the company's website naturally serves as a first point of contact for those interested seeking answers to general questions. The following topics will be offered online:

- Overview of the services offered by **ChemKit**;
- Presentation of the number of successful projects (possibly in anonymised form);
- Offering a contact point seven days a week around the clock.

5.2 Product – our service

We see our service as an all-inclusive service package that consists of consulting prior to decision-making, implementation support and post-changeover monitoring and controlling. The customer profits most of all from the interdisciplinary team of consultants, who not only are able to answer technical questions but can also elaborate on business and legal considerations. Due to the information events, a great deal of interest from users is expected. The employees of ChemKit will visit the various companies and discuss the options for implementation individually and in detail. In this way, the feedback of these events can be measured and after a certain time a decision can be made whether these efforts should be intensified or replaced by other activities.

5.3 Price – the pricing model

ChemKit charges a flat fee of EUR 3,000 for a consultation that is not followed by a changeover to Chemical Leasing. Half of this fee is paid by the BMLFUW. If the company decides to implement the Chemical Leasing scheme, this fee is dropped and the customer is charged EUR 8,000 for the service provided. As the Ministry subsidises the consultation, the user is again only charged half the fee, i.e., EUR 4,000. This fee is based on the average person-days expended on such projects, however the user will still enjoy savings in the 1st year of implementation. In the following three years, ChemKit will receive a 20% (EUR 1,200) share of the savings. The total revenue for every concluded project therefore is EUR 11,600. The table below shows a forecast (likely case) of sales and revenues in the SME market. As of the 2nd year, ChemKit plans to acquire a number of large-scale projects (LSPs). These are included in the forecast (Fig. 20).

	SME projects consulting ∅ EUR 3,000	**SME projects completed ∅ EUR 11,600**	**Large-scale projects consulting ∅ EUR 6,000**	**Large scale projects completed ∅ EUR 91,000**	**Revenue**
1st year	9	26	0		EUR 235,000
2nd year	20	57	1	1	EUR 608,000
3rd year	50	150	2	2	EUR 1,584,000
4th year	59	174	2	2	EUR 2,007,000

Fig. 20. Likely case: Revenue and number of projects in the SME market during the first four years

A survey among general managers of Austrian SMEs carried out by **ChemKit** has shown that the environmental benefits of the new model can have a substantial influence on the readiness to change over to Chemical Leasing (one of the respondents would even be prepared to give away all savings due to the benefit for environmental protection).

Because of the various applications and methods of Chemical Leasing, the pricing method can be adapted as needed and can be flexibly restructured. The **ChemKit** survey of Austrian SMEs also showed that changeovers of this type must bring rapid successes in order to be convincing. Initial expenses that are too high would tend to scare away potential customers. The pricing model of **ChemKit** is therefore designed with longer-term sharing in the savings – what is known as profit-sharing. The underlying daily rate of the basic model is EUR 800 (EUR 100 per hour). This price is justifiable as the consulting costs are amortised by the resulting efficiency gains.

The basic model is the result of the calculated average values of the market and industry analysis, as well as well-founded experts' opinions.

Person-day: 8 hours at EUR 100 each
∅ Project support: 8–9 person-days
∅ Savings of one SME: EUR 6,000 per annum (Figs. 21–24)

This results in the following revenue progression for **ChemKit** as well as savings progression for the user (Fig. 25).

The user finances the cost of changeover from the savings made and is additionally subsidised by the BMLFUW **during the first year. As of the start of the fifth year, it retains 100% of its ongoing annual savings of EUR 6,000.**

During the first year, ChemKit receives the biggest share of the total service charge and skims off 20% of the effective savings (∅ EUR 1,200) in each of the three following years. This results in a revenue of EUR 11,600 for each completed project.

The same pricing system, albeit with different benchmarks, applies to Austrian **large-scale projects**. As there are no suitable data available on the already completed projects at a national and international level, these figures are based on careful estimates (estimated average potential savings: EUR 60,000 per year). Conclusions about potential savings and revenue progression have been drawn analogously to the SME area. The contact phases with the relevant money flows are as follows (Figs. 26–28).

The revenue progression is also similar to that of SMEs (Fig. 29).

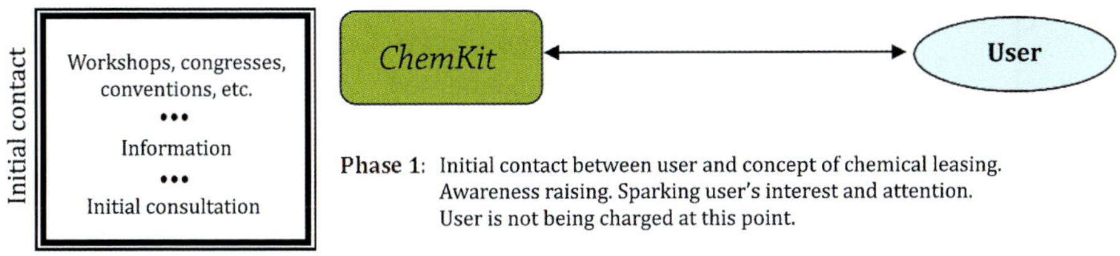

Fig. 21. 1st stage of implementation of the basic Chemical Leasing model

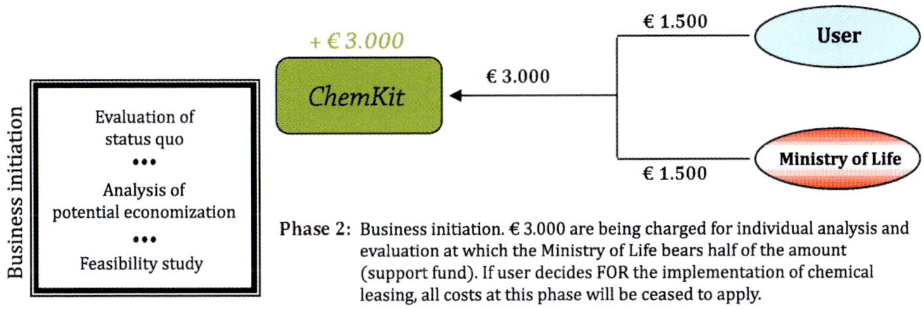

Fig. 22. 2nd stage of implementation of the basic Chemical Leasing model

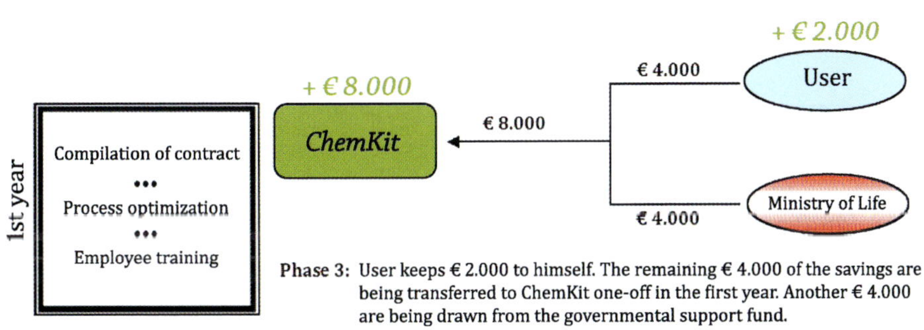

Fig. 23. 3rd stage of implementation of the basic Chemical Leasing model

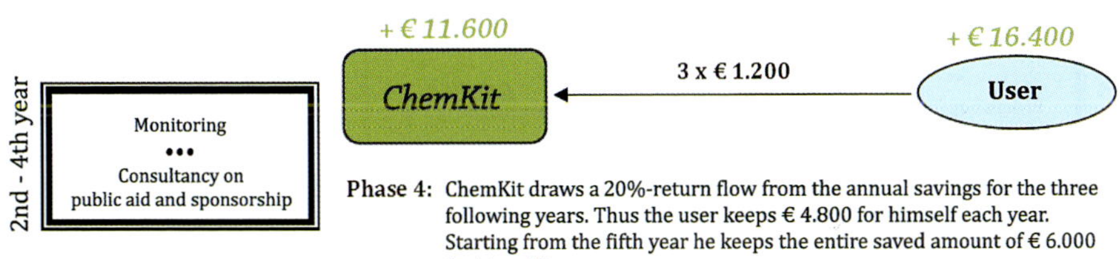

Fig. 24. 4th stage of implementation of the basic Chemical Leasing model

Fig. 25. Revenues of ChemKit and customer's savings of a base model project from year 1 through 5

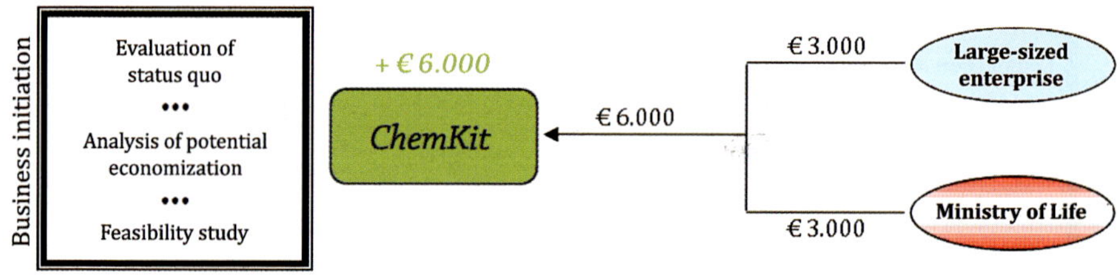

Fig. 26. Business initiation for large-scale projects

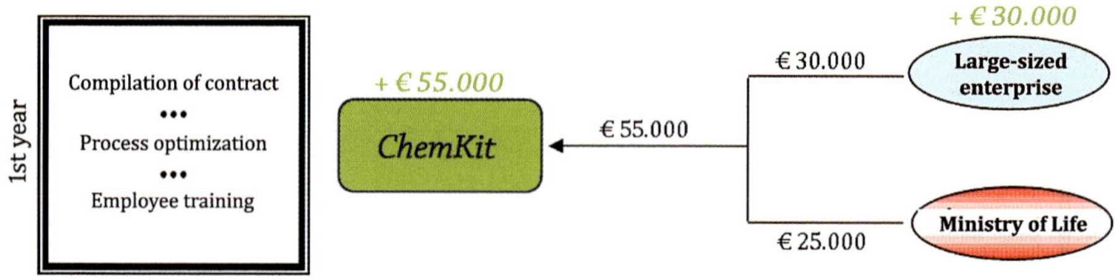

Fig. 27. Activities and money flows in the 1st year of implementation of a large-scale project

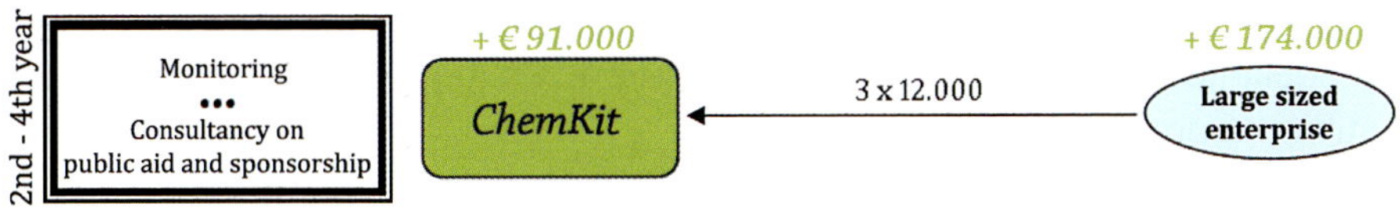

Fig. 28. Activities and money flows in years 2–4 of implementation of a large-scale project

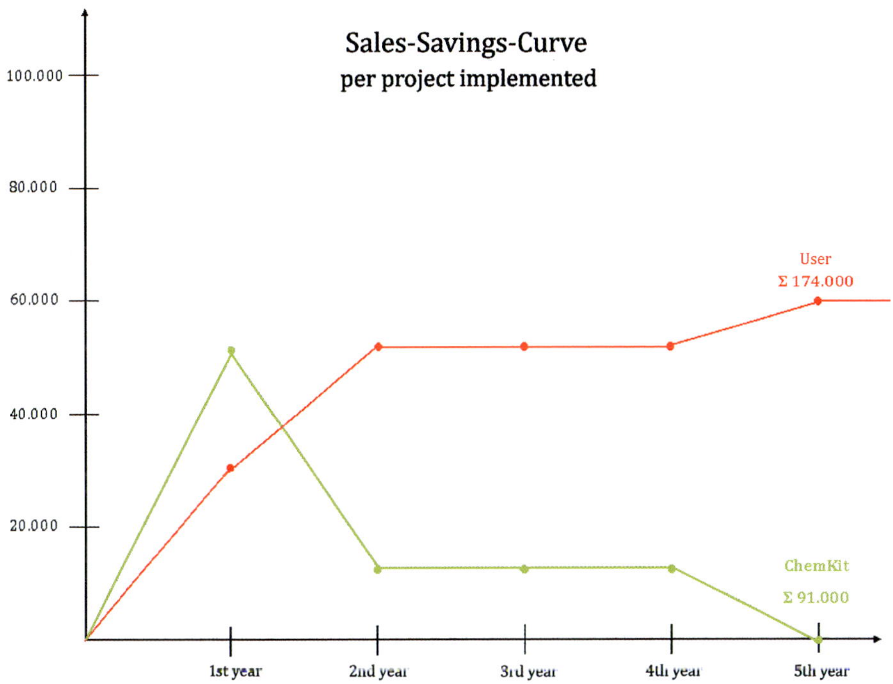

Fig. 29. Revenue progression of ChemKit and customer's savings of a large-scale project in years 1 through 5

6. Team of entrepreneurs and key positions

The founding team consists of three founding members. Aside from the technical-chemical and process-engineering expert knowledge, the general manager should also have a degree in economics or law in order to have the right to carry on a consulting business. To warrant reciprocal control of business operations, the other founding members have a right of co-determination for exceptional decisions. 40% of the company's nominal capital is contributed by the general manager and 30% by each of the two other founding members. For social insurance reasons and to optimise tax expenses, they will be given a full-time position at **ChemKit** GmbH.

The partners will collaborate with the person in charge of processes in the user companies and advise them on inter-company implementation and optimisation of the chemical and technical processes. Furthermore, they will co-ordinate the business aspects of Chemical Leasing projects, such as the logistical optimisation and profit-sharing between chemical supplier and chemical user.

All three persons should therefore have comprehensive expert knowledge of chemical technology, process engineering and economics. It is expected from them to have a degree in one of these three areas and a number of years experience in chemical process optimisation, process engineering, as well as specific industry and market knowledge. The business knowledge of the founding members must be of a sufficient level for them to be able to carry out the business aspects of Chemical Leasing projects in a satisfactory manner. Leadership experience and experience in the consulting field would be an advantage.

Their personal strengths should include negotiating skills, communication talent, charisma and a convincing power when dealing with customers. Any existing business contacts to potential customers would be desirable, as these will make entry into the market segment significantly easier. Close co-operation with experts of the BMLFUW, UNIDO and the certifying authorities (TÜV Süd, ISO) is already ascertained, since they wish to make Chemical Leasing commonplace in Austria and will offer the necessary standardisations to allow for a rapid diffusion of the model.

The **ChemKit** personnel planning will be as follows (Fig. 30).

The team competencies will furthermore be expanded and complemented by external experts.

An external commercial law specialist will be consulted for legal issues. The latter should already have relevant experience in chemicals laws. His expert knowledge is required for contract design, to be able to regulate the profit-sharing and risk distribution arrangements between the parties.

Year	2008	2009	2010
Best case	1 additional employee	6 additional employees	1 additional employee
Likely case		4 additional employees	1 additional employee
Worst case		2 additional employees	

Fig. 30. Personnel planning: Best, likely and worst case

A tax advisor will advise the team of founders on preparing the balance sheet and optimising tax.

7. Implementation plan

Based on the envisaged growth development, the following crucial milestones are defined for the next five years until ChemKit becomes established in the national and international consulting markets (Fig. 31).

	2007				2008	2009	2010	2011
	Q1	Q2	Q3	Q4				
Founding: ChemKit goes live!	■							
Market entry for SMEs (Austria): 1st project		■						
Completion of 10th project			■					
Break-even point reached				■				
Market entry for large companies (Austria)					■			
Completion of 50th project					■			
Completion of 100th project					■			
65% market share (Austria)						■		
EUR 1.5 mio in revenues						■		
EUR 1 mio in profits						■		
4 additional employees						■		
Completion of 150th project						■		
Negotiations for international co-operations; co-operation with international organisations							■	
First completed European/international projects								■

Fig. 31. Representation of business objectives/milestones

After the preparatory stages have been completed, **ChemKit** will officially be founded at the end of March 2007. This marks the commencement of our consulting activity. Up to the second quarter of 2008, the **ChemKit** team will focus on intensive, personal and competent consulting of small and medium enterprises in Austria; thereupon, the range of services will also be offered to Austrian large-scale companies.

The targeted market share (likely case) of **ChemKit** will already amount to an impressive 65% in the second year after founding. The main reason for this pleasant development is the (as yet) low level of competition in the Chemical Leasing consulting market. In the following years, it must be expected that market share will decline slightly.

Moreover, 2009 is the year in which the revenue threshold of EUR 1.5 mio will be reached and exceeded, which translates into a profit of EUR 1 mio. Project co-operation with large, small and medium-sized companies is profitable and is likely to lead to growing revenues and profits in the following years.

As the number of completed projects grows together with the circle of potential consulting customers (large, medium and small enterprises), it is likely that we will need to take on more employees in 2009. The extended **ChemKit** team will offer enough personnel resources to take on entry into the European consulting market as of 2010. The initiation and actual implementation of Europe-wide Chemical Leasing co-operations is an important and substantial step for **ChemKit's** development into an internationally accepted consultancy.

Achieving the break-even point in the 3rd quarter of 2007 is an important milestone for **ChemKit**. Exceeding the break-even point will show that the business policy and range of services finds acceptance in the domestic market and that **ChemKit** is already able to carry out enough projects to cover all costs and earn additional profits.

As a final step of development, **ChemKit** will establish contacts to international organisations in late 2010 and early 2011. In this connection, setting up relations with UNIDO is of special significance, since that organisation is already supervising numerous Chemical Leasing projects in the international markets and therefore has valuable experience and contacts. By co-operating with these institutions, **ChemKit** hopes to complete the implementation of its first international projects by 2011, opening the door to more Chemical Leasing co-operations around the world.

8. SWOT analysis (Fig. 32)

Rather than being just a consulting company, ChemKit – in co-operation with a chemical supplier – enables customers to outsource business processes that are not part of their core competencies. ChemKit is the one-stop supplier of all information necessary for the successful implementation of Chemical Leasing (technology, chemicals, business, law) and focuses its consulting activities exclusively on Chemical Leasing. This allows for a clear and unique positioning in the consulting market. ChemKit *gains a further mark of uniqueness*

Strengths	**Weaknesses**
✓ Coverage of all areas of relevance to contract design between chemical supplier and chemical user (technology, chemicals, business, legal) ✓ Unique co-operation and profit-sharing models ✓ Viability and profitability of the business model is proved by pilot projects ✓ Focus on Chemical Leasing allows ChemKit to clearly position itself in the consulting industry	✓ Not established in the market at first; lack of reference customers and no pilot projects of its own ✓ Low revenues in the early stages ✓ Word-of-mouth advertising difficult because customers wish to keep the competitive advantage offered by the Chemical Leasing strategy a secret
Opportunities	**Threats**
✓ Macro-trend towards sustainable environmental protection and social responsibility opens up, e. g., opportunities for marketing reaching companies participating in Chemical Leasing ✓ Increasing tightening of European chemicals policy (REACH) ✓ Public subsidies of consulting costs substantially reduce the changeover costs for the customer ✓ Public organisations promote Chemical Leasing (BMLFUW, UNIDO)	✓ Competitors could copy the consulting model ✓ Chemical suppliers offer Chemical Leasing schemes themselves ✓ Legal requirements for handling chemicals are greatly loosened ✓ Companies using chemicals transfer their production to low-wage countries ✓ Subsidisation for Chemical Leasing by the Federal Ministry is ended or the requirements for eligibility change to ChemKit's disadvantage ✓ Technological innovations are made that eliminate the use of chemicals in many processes

Fig. 32. SWOT analysis

vis-à-vis the competition through its refined profit-sharing offers to bring about win-win situations, and its close co-operation with the chemical supplier, which guarantees the highest level of efficiency and effectiveness of chemical use.

As a start-up company, ChemKit does not have a portfolio of pilot projects at this time. Here the Austrian consulting market offers an attractive niche to allow the company to become established. The viability and profitability of the business idea can be proved by numerous pilot projects, which will serve as a trust-building reference for potential customers.

Sustainability and efficient use of resources are THE keywords of our time and reflect the increased environmental conscience of society. This development provides producing companies with an incentive to engage in good governance and strongly promotes the implementation of Chemical Leasing. This trend is also followed in the EU chemicals policy – REACH is a coherent system for registering, evaluating and authorising chemicals, and will come into effect in mid-2007. For chemical suppliers, the co-operation with **ChemKit** sets up prerequisites to meet the REACH requirements. Chemical users are released from all and any obligations under the REACH policy. **ChemKit** can use the global and international developments to become established in the Austrian market and for its continued growth. Renowned organisations such as the Austrian Federal Ministry of Agriculture, Forestry, Environment and Water Management (BMLFUW) are highly enthusiastic about supporting the new model. Thus the attractive subsidy schemes of the BMLFUW lower the financial hurdle of companies who wish to use the services offered by **ChemKit**.

As the business model can generally be imitated, it is necessary that the market position be strengthened immediately after the business has been founded. All resources should be used for the customer acquisition process. **ChemKit** must use its pioneering position to become established as THE leading consultant in the Chemical Leasing field and thus to raise barriers to entry for potential competitors. Another possibility is that the chemical suppliers themselves may try to offer Chemical Leasing by themselves, without co-operating with **ChemKit**. As was clearly defined in talks with Dr. Schwager (UNIDO), Chemical Leasing projects absolutely require an "impartial third party" who ensures that the technical requirements for processes are met (e.g., the optimum processing time for metal parts to be coated with powder) and supervises the observance of the agreed quality standards, so that neither the chemical supplier nor the chemical user can take advantage of the project partner. **ChemKit** will make sure to communicate the importance of an impartial third party to the involved parties emphatically in its marketing activities.

Other threats to the profitable realisation of **ChemKit** would be if the subsidies offered by the BMLFUW were to be withdrawn, or if the European chemicals policy were to make a fundamental change of direction. One example here would be that, as a result of the competition with transition countries for attractiveness as a business location, the requirements for handling chemicals might be loosened substantially. Another possibility would be that Austrian companies which use chemicals and for which Chemical Leasing would be a viable option shift their production to low-wage countries. However, by the time such significant changes of the political and legal climate and economic structure will take effect, **ChemKit** will have expanded its activities into the international arena. Moreover, the pilot projects successfully carried out by UNIDO in various transition countries prove that the attraction of the Chemical Leasing model not only comes from its benefit to the environment and compliance with legal requirements, but especially from its financial benefits.

9. Financial plan

9.1 Expenses

Since **ChemKit** is a pure consulting company, the financial plan has a relatively simple structure. Apart from start-up costs (company registration, notary public, fees and so on), the expenses to be incurred in the first year will only be one-off payments for office and business equipment and running costs. Great importance will be attached especially to the marketing expenses, which will be particularly high in the first months. This includes, among others, information events, trade fair appearances, workshops, printed matter (calling cards, prospects, advertising material), website production, etc. The company's own capital of EUR 35,000 is intended to be fully used up by the partners.

The following table will show the most cost-intensive items for the year of founding (2007) (Fig. 33).

Human resources costs

For the founding of the company, the partners are willing to forgo a high salary at the beginning. In the first year, each partner will earn EUR 21,000 net (this equates to a monthly income of EUR 1,500). Only when the company's profits rise and the company grows will these salaries grow. In the fourth year (2010), each partner will earn a monthly salary of EUR 3,500 net. For tasks that are not

Human resource costs	EUR 66,840.00
Expenses for motor vehicles (first payment + monthly instalments)	EUR 38,388.00
Rent for business premises (incl. operating costs and deposit)	EUR 13,700.00
Marketing	EUR 11,500.00
Travel expenses	EUR 6,000.00

Fig. 33. Biggest investment items in the year of founding

3 partners	EUR 54,000.00
External service providers (tax and legal consultants)	EUR 7,840.00
eBüro	EUR 5,000.00

Fig. 34. Breakdown of human resources costs

fully within the competencies of the team of founders, external service providers such as tax advisors and a solicitor, whose fees are calculated on the basis of hours worked, will be consulted. Instead of setting up a secretary's office, **ChemKit** will employ the services of an office service provider such as eBüro AG[12] for 16 months starting March 2007, since the administrative activities will have a very simple structure at that time. This ensures round-the-clock availability by telephone. The office service provider will take calls, coordinate appointments and pass on information. The progressively designed invoicing model of such an office service provider is seen as particularly attractive in the year of founding if compared to the high fixed costs of a secretary's office with full-time staff. Full-time secretarial staff will only be employed in the second half of 2008, as **ChemKit** will continue to grow successfully (Fig. 34).

9.2 Revenue planning (Fig. 35)

The payments received are calculated on the basis of the calculation model shown in the marketing section. They comprise revenues from consulting and project implementation both for the SME and large-scale company segment (Fig. 36).

12 www.ebuero.de

	SME projects consulting ∅ EUR 3,000	SME projects completed ∅ EUR 11,600	Large-scale projects consulting ∅ EUR 6,000	Large scale projects completed ∅ EUR 91,000	Revenue
1st year	9	26	0	0	EUR 235,000
2nd year	20	57	1	1	EUR 608,000
3rd year	50	150	2	2	EUR 1,584,000
4th year	59	174	2	2	EUR 2,007,000
Due to the accounting method used, payments from the profit-sharing scheme will only be received in the subsequent years.					

Fig. 35. Planning of revenues and number of projects 2007–2010

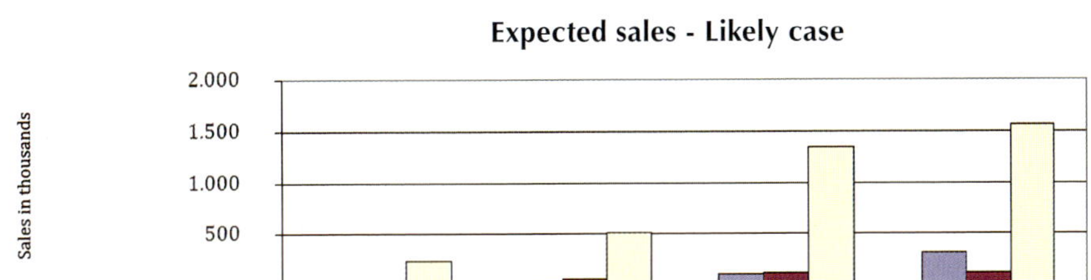

	2007	2008	2009	2010
Return flows of initiated projects		31	112	316
Large-sized enterprises		61	122	122
SME	235	516	1.350	1.569

Fig. 36. Expected revenues for ChemKit

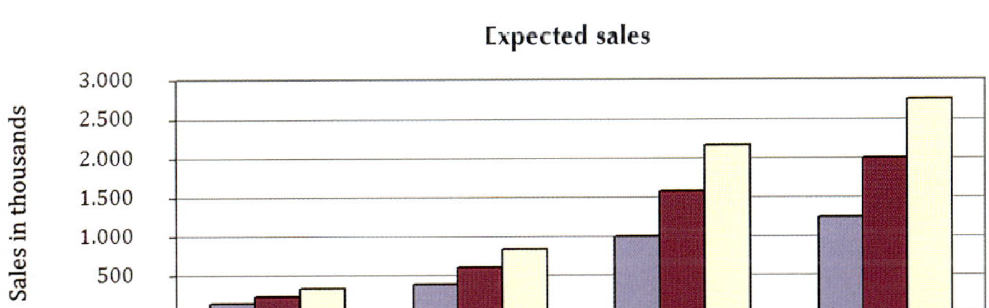

	2007	2008	2009	2010
Worst Case	143	388	1.004	1.253
Likely Case	235	608	1.584	2.007
Best Case	340	836	2.167	2.751

Fig. 37. Best, likely and worst case scenario of the expected revenue progression 2007–2010

Scenario management

In the following, three scenarios are compared according to their revenue expectations. The basis for the calculation of the variance coefficients is the theory of the diffusion curve[13] of innovations. It is assumed that the Austrian market will be satiated by 50% after 10 years. This estimate is based on the analogous market of "Responsible Care". Expected market shares have been used for the calculation of the variance coefficients of the Worst and Best Case scenarios, which assume a 30% and a 70% satiation, respectively (Fig. 37).

References

Sources

Bundesverband Deutscher Unternehmensberater. Facts & Figures zum Beratermarkt 2005/2006 (pdf)

FEACO – European Federation of Management Consultancies Associations (2004) Survey of the European Management Consultancy Market

Jakl T, Joas R, Nolte R, Schott R, Windsperger A (2003) Chemikalien-Leasing. Springer, Vienna

KMU Forschung Österreich (PDF 3. Quartal 2006)

Kotler P, Bliemel F (2001) Marketing-management, 10th edn. Schaeffer-Poeschel, Stuttgart

Martinuzzi A (2005) Erfolg im Öko-Consulting. Hampp, Mering

Moser F (2005) Chemikalien Leasing und die Verringerung negativer externer Effekte: Der "Property Rights" Ansatz. Diplomarbeit, Geneva

Schriftenreihe des BLMFUW (2003) Chemikalien-Leasing: Modell für eine nachhaltige Stoffwirtschaft", 2nd edn, Bd 13/2002

Internet sources

http://automotive.henkel.com/Services/Chemical+Management/Overview/Benefits.htm (last access: 18 January 2007)

http://reach-info.de/ (last access: 18 January 2007)

http://www.achema.de/ (last access: 18 January 2007)

http://www.bipro.de/ (last access: 18 January 2007)

http://www.chemicalstrategies.org/ (last access: 18 January 2007)

http://www.cmsforum.org/index.html (last access: 18 January 2007)

http://www.ebuero.de (last access: 18 January 2007)

http://www.fcio.at/ (last access: 18 January 2007)

http://www.feaco.org/content/content2.php?CatID=148&NewsID=406 (last access, 18 January 2007)

http://www.gm.com/company/gmability/sustainability/reports/02/000_navigation/030_printable/env.pdf (last access: 18 January 2007)

http://www.haastcm.com/ (last access: 18 January 2007)

http://www.umweltbundesamt.at/umweltschutz/chemikalien/reach/ (last access: 18 January 2007)

http://www.vienna-tec.at/ (last access: 15 January 2007)

13 Diffusion curve of innovations (for details, see Appendix)

Panel of experts

Dr. Reinhard Joas

General Manager of BIPRO GmbH – Consulting company for integrated problem solution
e-mail: reinhard.joas@bipro.de
Type of cooperation: expert interview in Munich (9 November 2006), e-mail correspondence

Dr. Thomas Jakl

Head of Department V/2: Product-related Environmental Protection and Chemicals Policy
e-mail: thomas.jakl@lebensministerium.at
Type of cooperation: expert interview in Reichenau (13 October 2006), meeting at the Austrian Federal Ministry of Agriculture, Forestry, Environment and Water Management (BMLFUW) (2 November 2006), telephone conversations, e-mail correspondence

Mag. Walter Beyer

Owner of BUK Beyer Umwelt + Kommunikation
e-mail: office@beyer.at
Type of cooperation: Expert interview at BUK Beyer, Vienna (17 November 2006)

Dr. Petra Schwager

Branch: Energy and Cleaner Production UNIDO
e-mail: p.schwager@unido.org
Type of cooperation: Expert interview at UNIDO Headquarters, Vienna (14 November 2006), e-mail correspondence

Dipl.-Kfm. Steffen Säcker

Business Development Manager SAFECHEM Europe GmbH, Düsseldorf
e-mail: ssaecker@dow.com
Type of cooperation: telephone conversation (8 November 2006)

Chapter 13

Chemical Leasing and Regulatory Approaches in Chemicals Policy

How Chemical Leasing Paves the Way to REACH-compatibility

Thomas Jakl

The process of redesigning the European Union's Chemical Policy was triggered by several factors. A large number of man-made chemicals has been released into the environment, is still being and will inevitably be released in the near future. There are several cases of chemicals which had been considered to be safe but which due to their persistent nature manifested environmental harm in areas outside of their initial assessment.

On the one hand chemicals were being traced in environmental media as well as in products in unexpected dimensions both in terms of quality as well as in terms of quantity. On the other hand effects like "endocrine disruption" showed that also extremely low concentrations (like ppt, parts per trillion; equals to the sugar concentration achieved if one lump of sugar would be dissolved in lake Constance) could lead to adverse effects. Moreover the current instruments of chemicals policy could – at least to a great extent – not cope with the political challenges.

A new model has been devised to specifically meet these political challenges of Europe's environmental policy with regard to chemical products as included in the 6th Environment Action Programme of the EU, which stipulates the following requirements (literature quote):

"Chemicals that are dangerous should be replaced by safer chemicals or safer alternative technologies not entailing the use of chemicals, with the aim of reducing risks to man and the environment."

The Action Programme also demands to place "the responsibility on manufacturers, importers and downstream users for generating knowledge about all chemicals (duty of care) and assessing risks of their use, including in products, as well as recovery and disposal."

It finally commits the European Union to aim at achieving "within *one generation (2020)* that chemicals are only produced and used in ways that

do not lead to a significant negative impact on health and the environment ...".

The EU's future Chemicals Policy, of which the REACH system represents the centrepiece, is the tool for achieving these goals. REACH is based on the principle of documentation, evaluation and minimisation of risks resulting from chemicals – which is above all to be implemented by the industry itself.

The REACH – philosophy in a nutshell

In particular the *reversal of the "burden of proof" will* be a key component of the REACH system leading to *a "no data – no market"* concept obliging producers or importers of substances to deliver a certain amount of documentation regarding properties of chemicals and possible risks during their applications as a precondition for market access. This implies a real shift of paradigm: Although the current legislation also was based to a certain extent on the producers' responsibility – the actual burden of proving a certain risk and of demanding risk-appropriate management measures however was with the authorities. REACH establishes a single, integrated system for Registration, Evaluation and Authorisation of CHemicals. It is requiring companies that produce and import chemicals to assess the risks arising from their use and to take the necessary measures to appropriately manage any risk they identify.

Under the same political emphasis it was ensured that the *"precautionary principle"* is the second cornerstone of the new legislation:

Concerning potential negative impacts of chemicals, policy makers are challenged to decide whether to take preventive measures or not. Especially in cases of blurred scientific evidence, applying precaution is *the* political and legal answer for decision-makers to deal with potential risks in order to avoid being late.

1. In 2001 the Austrian Federal Ministry of Agriculture, Forestry, Environment and Water Management (BMLFUW) hosted the scientific conference "The Role of Precaution in Chemicals Policy" (Hafner et al., 2002).

The chairs of that conference concluded inter alia:

- "Lack of certainty – as an agreed precondition for precautionary action – could lead to phasing out of substances or uses but could at the same time hamper substitution by alternatives because their properties are often even less well documented. There was broad consensus that pre-marketing measures ensuring the availability of an adequate set of information were seen as the prerequisite for overcoming this deadlock.

- Ideally, the Precautionary Principle is incorporated into a legislative context that sees to generate adequate data prior to production and marketing of a chemical or a product.“

This principle is fundamental to numerous multilateral agreements and treaties. As mentioned in the conclusions quoted above, in order to make Precaution operational, pre-marketing measures ensuring the availability of an adequate set of information on the properties of chemicals are essential. Thus in order to be "safe instead of sorry" REACH establishes not only the "no data no market" approach but also an authorisation regime for substances of very high concern as they are representing serious hazards to human health and the environment. Particular inherent properties are directly linked to risk reduction measures – an unprecedented application of the Precautionary Principle.

The new REACH system is setting high standards for the protection of health and the environment while aiming at safeguarding the competitiveness of enterprises and improving the potential for product innovation. This balance would be to the long-term benefit of chemicals manufacturers, importers, users, small and medium sized enterprises, consumers and for health and the environment.

REACH as a business perspective

Once these new conditions and circumstances will be established by the REACH-System, the conventional paradigm stipulating that chemicals are just sold by one side and purchased by the other without any further exchange of information, cannot survive. The responsibilities are too interwoven and deep-seated so that the importance of the relationship between manufacturer and user of chemicals can no longer be judged by commercial indicators only.

In co-operation with the OECD the Austrian Federal Ministry of Agriculture, Forestry, Environment and Water Management (BMLFUW) hosted the international conference „Experiences and Perspectives of Service-oriented Strategies in the Chemicals Industry and Related Areas“ in Vienna in November 2003. The conference concluded:

"11. All these new service-oriented chemical business models require a close co-operation between the provider and the user of the chemical. Therefore, the potential of these business models has also to be seen in connection with the new EU Chemicals Policy (REACH), which will require a new relationship between provider and user and the conventional paradigm „supplier here – customer there" will hardly be crowned with commercial success."

In the framework of Europe's future Chemicals Policy and due to the obligation to document environmental and health hazards associated with chemical substances, the *supplier-customer relationship* will be compelled to acquire an entirely new quality. The REACH system is in a way going to mandate an intensified dialogue between producers and users of chemicals in both directions of the supply chain:

On the one hand the *supplier* will have to document potential hazards of chemicals as well as risks occurring during their use and application. On the other hand also the *user* will be challenged to make available new experiences, gained during the handling and use of the chemicals, to its partners and to the system itself. These new obligations will require not only reasonable efforts by companies, but will also require a new culture in terms of information exchange, communication and cooperation.

REACH and Chemical Leasing share the same philosophy

Both approaches are mutually supportive as they both stimulate the development of rules for „sharing". Within REACH test costs will have to be shared between companies registering the same substance. A specific guidance document is being developed in that regard by the European institutions.

Companies will have to share responsibility as to the documentation of the properties of a chemical substance as well as with regard to the risks that might occur during its application. Also a great deal of information will have to be shared between companies representing various stages of the supply chain as applicants depend on up-stream documentation to be able to fulfil their obligation to document risks and applicable risk reduction measures for their specific use. This culture of sharing, which might be a new cultural element in the relationship of business partners, is also a prerequisite for success in service-oriented business models as they equivalently depend on a high degree of openness and trust between the partners involved.

The agreement, partners will have to develop as a basis for their co-operation, also requires mechanisms and procedures clarifying responsibility and liability patterns with regard to the performance and quality of their common business activity applying Chemical Leasing.

The Registration phase of REACH and the Principles of Chemical Leasing business models have additional central aspects in common. They both involve different stages of the supply chain – as producers and applicants are challenged – and both approaches are life-cycle oriented either through their documentation requirements addressing phases of production, use or disposal or through their integration of the corresponding partners managing those life-cycle stages within the business model.

Chemicals are to be handled with care both in the REACH world as well as within applications of Chemicals Leasing. Care in that context in particular implies that chemicals and their applications are not only monitored but also managed with maximum accuracy. REACH represents the regulatory driver for this attitude aiming at protecting human health and the environment ("Duty of care" in the REACH Regulation Recital 16, Art. 1 para 3; *monitoring requirements* Art 14.6. for manufacturers; Art 37.5. down stream users) whereas **Chemical Leasing** is additionally driving it from the economic point of view since resource efficiency simply increases profit.

Chemical Leasing and the "adequate control" route within the authorisation regime – a delicate balance

Chemical Leasing is *the* tool to demonstrate "adequate control" (Article 58), a set of parameters which have to be fulfilled in order to qualify for the authorisation of usage. As outlined above only substances of very high concern will have to be treated under the REACH authorisation regime and it is far from being my intention to recommend "Chemical Leasing" as a means to prolong the use of such extremely hazardous substances. However as a matter of fact, experience shows that chemicals are managed excellently within Chemical Leasing applications. In case the use of substances of very high concern is inevitable in certain applications, adequate control will thus be achieved, since "handling with care" is the core element of such Chemical Leasing service-oriented business models. In view of the reasons mentioned above there is a certain need to assure that the substitution of extremely hazardous substances by alternatives is an underlying principle also in Chemical Leasing applications. I would like to draw your attention to chapter 6 lining out "quality assurance" activities in that regard.

Chemical Leasing is the ideal business environment to identify and apply the *use and exposure category* concept in particular within the Chemical Safety Report – jointly by suppliers and users. Specifying the relevant use and exposure category within the REACH system, together with qualifying the risks possibly arising will build upon the assessment already performed during the establishment of the specific Chemical Leasing model. Also the development and application of appropriate risk management measures will be part of that process which therefore will highly likely render the core elements necessary to apply the exposure category concept within REACH.

REACH is going to mandate along the supply chain information exchange, monitoring procedures, patterns for sharing and co-operation as well as documentation and assessment procedures. Recital 17 of the REACH Regulation reads as follows:

"All available and relevant information on substances on their own, in preparations and in articles should be collected to assist in identifying hazardous properties, and recommendations about risk management measures should systematically be conveyed through supply chains, as reasonably necessary, to prevent adverse effects on human health and the environment. In addition, communication of technical advice to support risk management should be encouraged in the supply chain, where appropriate."

In judging and quantifying the efforts the REACH system will require, the positive effects for the industry's competitiveness must not be neglected. On account of the future legal requirements an intimate knowledge of the chemicals' features and the effects during their life cycle can become a factor that might positively influence competition.

Chemical Leasing is making use of REACH structures and is turning them into economic advantages while at the same time catalysing REACH compliance

The REACH system has the potential to intensify cooperation and to intensify exchange of experiences. The information generated within the REACH system serve as precious basis for know-how development in service-oriented strategies. In the framework of Europe's future Chemicals Policy and due to the obligation to document environmental and health hazards associated with chemical substances, the supplier-customer relationship will be compelled to acquire an entirely new quality. On account of the EU's legal requirements with regard to risk assessment, an intimate knowledge of the product features and the processing will be a factor that will positively influence competition. The more useful the data on the supplier's substances prove for the user, the less efforts need to be invested in risk assessment. The future undoubtedly belongs to business models that entail an intensive dialogue, cooperation and the bundling of creative potentials – and here Chemical Leasing takes on a leading role.

It is up to the implementation of the REACH system to make sure, that this intensive form of cooperation will mobilise synergies, which – contrary to prevailing apprehension – will give Europe a pronounced edge as a location for the chemicals industry.

Conclusion

New approaches towards chemicals policy and chemical management were developed during the last years on EU level as well as worldwide, initiating a shift in paradigm. Knowledge about the chemicals' properties and communication about the risks that might arise during their application are the pillars for the new attitude required for chemical business. Chemical Leasing opens a window of opportunity for turning new obligations, new responsibilities and new information flows into successful business strategies.

References

DECISION No 1600/2002/EC of the European Parliament and of the Council of 22 July 2002 laying down the Sixth Community Environment Action Programme, 10. 9. 2002 Official Journal of the European Communities L 242/1

REGULATION (EC) No 1907/2006 of the European Parliament and of the Council of 18 December 2006 concerning the Registration, Evaluation, Authorisation and Restriction of Chemicals (REACH), establishing a European Chemicals Agency, amending Directive 1999/45/EC and repealing Council Regulation (EEC) No 793/93 and Commission Regulation (EC) No 1488/94 as well as Council Directive 76/769/EEC and Commission Directives 91/155/EEC, 93/67/EEC, 93/105/EC and 2000/21/EC, 29. 5. 2007 Official Journal of the European Union L 136/3

Hafner G, Jakl T, Loibl G (2002) Conference conclusions by the chairs. In: Freytag E, Jakl T, Loibl G, Wittmann M (eds) The role of precaution in chemicals policy. Favorita Papers, Diplomatic Academy of Vienna

OECD, Organisation for Economic Co-operation and Development, Environment Directorate, 13-Dec-2004, ENV/JM/MONO(2004)28/PART1, Series on Risk Management No 17, "Workshop on experiences and perspectives of service-oriented strategies in the chemicals industry and related areas", Vienna, Austria, 13–14 November 2003, Part I: Summary and conclusions

Chapter 14

Conclusion

14.1 Lessons Learned – The Way Forward

Thomas Jakl and Petra Schwager

Service-based business models in chemicals management of course existed in various areas before the term "Chemical Leasing" was coined. The merit of our – UNIDO and the Austrian Federal Ministry of Agriculture, Forestry, Environment and Water Management (BMLFUW) – initiative is that we were able to document that this idea and approach can be successfully implemented within an astonishingly broad range of applications of chemical products and that there are several best practice examples showing how Chemical Leasing is really becoming a "win-win" story. In particular, we have observed the outstanding potential for economic and environmental improvements that can be achieved by the application of Chemical Leasing and Cleaner Production. Demonstration projects implemented in a number of different sectors in close cooperation with UNIDO's National Cleaner Production Centres have shown that this potential might even be higher in developing countries and countries with economies in transition. "To base chemical business transactions on units describing the performance of products is a promising option also for small and medium-sized enterprises all over the globe" is another possible key message based on our findings.

When applying Chemical Leasing business models, the producer does not sell the chemicals but rather the know-how associated with them. This relates to optimising the amounts of chemicals used for the task required, conditions of use, recycling concepts and disposal. Chemical Leasing thus supports and further enhances the positive impact of the Cleaner Production approach. According to the definition of Cleaner Production, it is the continuous application of an integrated preventive environmental strategy to processes, products and services to increase overall efficiency, and reduce risks to humans and the environment. Cleaner Production can be applied to processes used in any industry, to products themselves and to various services provided in society. It refers to a mentality of how goods and services are produced with a minimum of environmental impact under present technological and economic limits. Combining Cleaner Production and Chemical Leasing is therefore a clear win-win strategy to achieve sustainable industrial development: It protects the environment, the consumer and the worker while improving industrial efficiency and competitiveness.

After half a decade of experience with regard to the implementation and evaluation of Chemical Leasing projects, some trends can be clearly recognised permitting a comprehensive analysis. What has been postulated in a number of publications, presentations and statements can now be judged on the basis of well-documented facts and figures:

- In terms of typology and categorisation, in many cases the supplier of the chemical substances retains the ownership of the chemical whereas the applicant usually retains the ownership of the application plant. So it seems that in general the core competences of the partners involved are maintained while they are embedded in an umbrella model based on a case-specific unit of payment.
- The consensual identification and control of the unit of payment together with an agreed approach towards profit sharing seemed to be the most time-consuming and sophisticated steps during the development of the model. Instruments such as UNIDO's Toolkit or quality assurance systems will be supportive in managing that critical and important stage.
- Once established, Chemical Leasing approaches satisfied all the partners involved, who unanimously declared that implementation was successful and that they do not at all regret having carried it out and that its merits are obvious, particularly for small and medium-sized enterprises. In this context, the coordinating activities of independent entities, such as the UNIDO National Cleaner Production Centres (NCPCs), are considered to be important key factors for success and also ensure adequate benefit sharing and proper monitoring of Chemical Leasing projects.
- The concrete application at plant level has furthermore demonstrated that a systematic approach based on the preventive Cleaner Production idea boosts the economic and environmental success of Chemical Leasing business models.
- It was confirmed that Chemical Leasing mobilises synergies while combining the know-how of producer and applicant of chemical products and establishing stable and profitable business relationships.
- Compared to business models run before Chemical Leasing was implemented, all cases showed remarkable increases in resource efficiency and environmental performance.
- From a political point of view, one of the key findings is that Chemical Leasing does have obvious advantages within modern legal frameworks

such as the REACH system, in which responsibilities concerning risk assessment and risk management are distributed throughout the supply chain.

In addition, the evaluation of the various projects showed that resource efficiency and process optimisation are the key drives within the model, while the ecological optimisation of the chemical substances involved is not a matter of course. This is why environmental policy has its place in safeguarding the proliferation of Chemical Leasing, as its ecological profile (with regard to the substances used) is not invulnerable. The continuing political support of this business model therefore has to be equipped with tools like quality-assurance schemes or other instruments such as auditing or eco-labelling schemes in order to ensure maximum environmental benefit.

It is Austria's political goal to establish Chemical Leasing as a prominent element within national, regional and global initiatives, programmes and strategies aimed towards the implementation of sustainable chemicals management. Not because this is a "one fits all" approach, but because there is plenty of evidence documenting the fact that its implementation is a promising alternative, which – in a multitude of scenarios – has proved to be at least worth considering. In those cases where Chemical Leasing was actually implemented, the obvious economic and ecological success permitted the partners involved to continue operating within the model.

As far as political institutions are concerned, their status and role as supporters of the idea has to be designed carefully. Chemical Leasing is a concept that has to be developed and implemented by companies on a case by case basis. It is they who are responsible and it is they who bear the possible risks. So it can only be the task of political institutions to signal their sympathy and to offer support. This could be financial and logistic support on the one hand and support in terms of information and guidance on the other. According to our experience, theoretical information of a high quality is of course a prerequisite for successful implementation – however, the more authentic and experience-based the documentation is, the greater the credibility and relevance it has for companies.

As far as UNIDO is concerned, encouraging innovative concepts to ensure progressive and environmentally sound development constitutes an integral element of its Cleaner Production strategy. Being an industrial development organisation it is the aim of UNIDO to promote the novel service-based Chemical Leasing models and support sustainable chemicals management in developing countries and countries with economies in transition. In this undertaking, UNIDO fosters national capacity building, the implementation of demonstration projects and the adaptation of the concept to national requirements. Developing tools and instruments to support the global implementation of efficient Chemical Leasing business models based on the preventive approach of Cleaner Production is another important issue of UNIDO's work.

It is far from exaggerating when we postulate that due to the experience gained so far, Chemical Leasing is becoming a worldwide perspective for sustainable development within chemicals management. Thus Austria will further increase its support of that idea. The catalogue of examples and evaluations contained in this book will pave the way to a broad implementation of Chemical Leasing while minimising the economic risks for companies. Here the UNIDO worldwide network of Cleaner Production Centres has the key for Chemical Leasing to go global.

14.2 Typology of the Case Studies Presented

Andreas Windsperger and Richard Tuschl

In this chapter, we try to classify the presented cases in order to facilitate comparing the different cases concerning their characteristics, their performance and their acceptance by users.

For the classification a typology is used, which has been worked out in a project of the Austrian Federal Ministry of Agriculture, Forestry, Environment and Water Management (BMLFUW) and is published in the book „Chemical Leasing".

As the previous chapters show, all the cases focus on providing the service instead of selling chemicals. But this principle can show different faces in practice, sometimes intermediary solutions resulting from increased customer care or manufacturer responsibility are used, while only the very serious ones are based on an actually service-oriented approach. The "unit for payment" turned out to be a fundamental issue for the scope of a service. Many contracting models, although called "service-models", only represent models of improved customer care. If the amount of the supplied product is the basis for the payment and if the ownership of the product changes, it is not Chemical Leasing, but a somehow expanded responsibility approach. But if the user is charged for the service, for the function of the used chemicals instead of paying for used material amounts, then this is a clear indication of a service model.

Such approaches sometimes represent a distinct risk for the contractor in such models, since misuse of a user can hardly be excluded without giving the user the impression that the service has been curtailed. A clear separation of responsibilities, ownership, operation, maintenance, etc., in the individual processes helps to avoid dissatisfaction. Service models should target at an almost complete takeover of responsibility not only for the chemical, but also for the application plant and the waste management or recycling equipment.

To look in detail on the presented cases the typology uses criteria which form a neutral picture of the character of the developed service system. They allow easy comparison between the cases and other existing business models, pointing out differences and looking for further potential improvements in order to develop target-models.

The classification of the models is based on the following criteria:

- who owns the substance
- who owns the plant

- who operates the plant
- who maintains the plant
- where is the plant situated.

The processes supply, use and waste management or recycling as well as the respective stakeholders are considered. The basis of payment (unit of charge) is used as an indication of the service character.

The characteristics above are described for each of the individual cases and presented as a table and a chart. Finally the characteristics of all the cases are summarised in a table and some conclusions are drawn.

To facilitate an overview of the characteristics of the individual models, a presentation as an X-Y chart has been chosen, which allows to show all the properties at a glance. On the Y-axis the processes are drawn, their location is visible on the X-axis. The processes are also indicated by symbols. The supplier (or supply, "S"), the user (or use "U") and the waste manager (or waste disposal "W") are indicated on both axes pointing out where which activity takes place. Colours indicate the ownership of the substance and the plant in the various phases. The colour of the symbol area denotes the owner of the substance, the colour of the symbol frame indicates the owner of the plant. Finally, a coloured dot in the symbol shows who operates the plant. The arrows between the symbols indicate the movement from one location to another while an eventual outer frame around the symbols shows the possibility of bundling plants at one single location (Fig. 1).

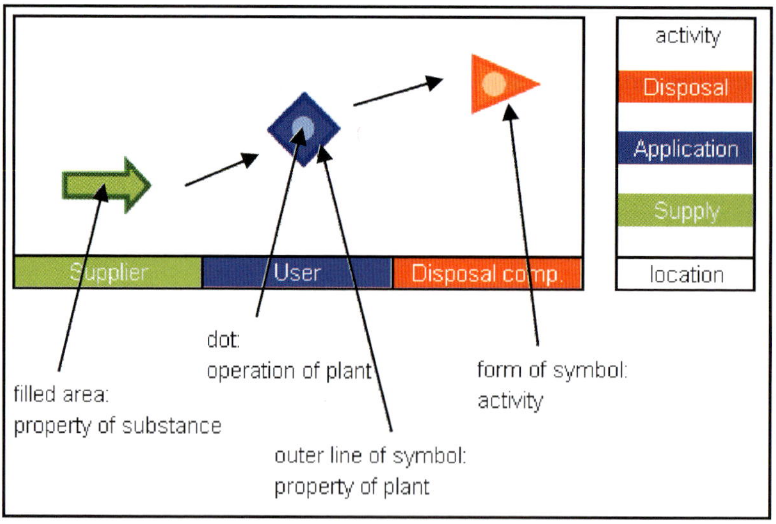

Fig. 1. The model

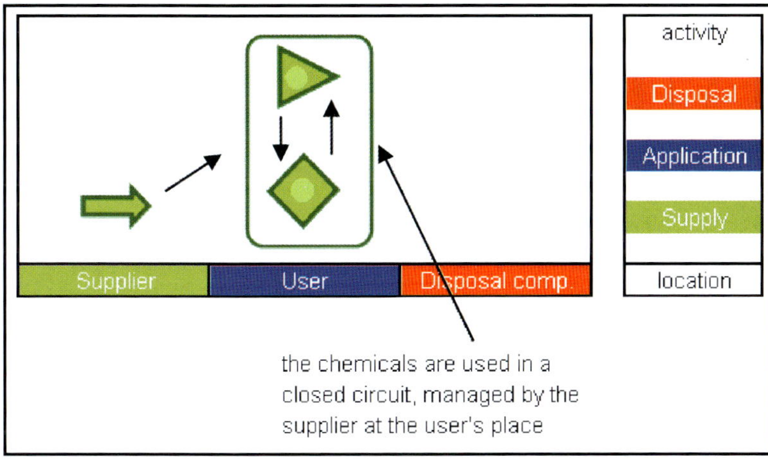

Fig. 2. Example for a "Total Care Model"

As an example for a high level service model the so called "Total Care Model" may serve, where the supplier owns the application and the recycling equipment and is responsible for its operation and maintenance, although the plant is situated at the user's premises (Fig. 2).

Description of the presented cases

Chemical Leasing in the field of de-lacquering
Tiefenbacher GmbH
Mepla – Alfit
Austria

Mepla – Alfit needs special hooks for the painting process of the different parts. An organic solvent mixture is used for internal de-lacquering and is delivered and also collected after use by Tiefenbacher. Since de-lacquering at an external plant is impossible due to organisational and economic reasons the know-how of the de-lacquering company was integrated into the process at Mepla – Alfit. The chemical management was outsourced to the de-lacquering company and deliverer of the solvent. So the application plant is in full responsibility of the user, while the delivery, know-how transfer, supervision and recycling are done by the service provider. At the beginning the basis of payment was the consumed quantity of chemicals. The goal is to find another basis with a high correlation to the function of the chemicals (Fig. 3).

	Chemical	Application equipment	Recycling Disposal
Owner	S	U	S
Location		U	S
Maintenance		U	S
Operation		U	S
Charging	Quantity		

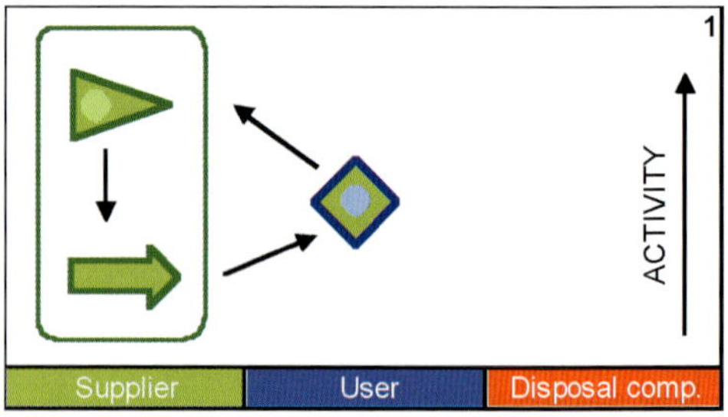

Fig. 3. Business case 1

Chemical Leasing in electrostatic powder coating
Akzo Nobel Powder Coatings S.A.E
ABB ARAB
Egypt

ABB ARAB is a subsidiary company of the international ABB group. The company is the leader in the market of manufacturing electric equipment, especially high and low voltage equipment. For components of the power equipment ABB is operating an electrostatic powder coating process, for which Akzo Nobel is supplying the electrostatic coating powder. Under the Chemical Leasing (ChL) business model, Akzo Nobel provides the coating powder, manages and supervises the powder coating process at ABB ARAB. The service provider is also responsible for the recycling process. The basis of payment is the coated area per month against a fixed price (Fig. 4).

	Chemical	Application equipment	Recycling Disposal
Owner	S	U	S
Location		U	S
Maintenance		U	S
Operation		S	S
Charging	Service		

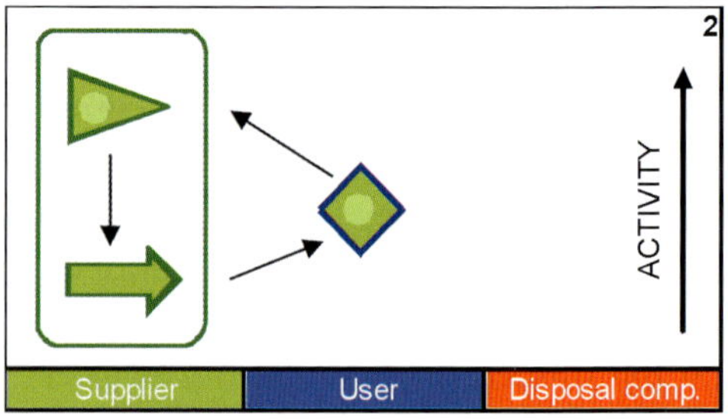

Fig. 4. Business case 2

	Chemical	Application equipment	Recycling Disposal
Owner	S	U	S
Location		U	S
Maintenance		U	S
Operation		U	S
Charging		Service	

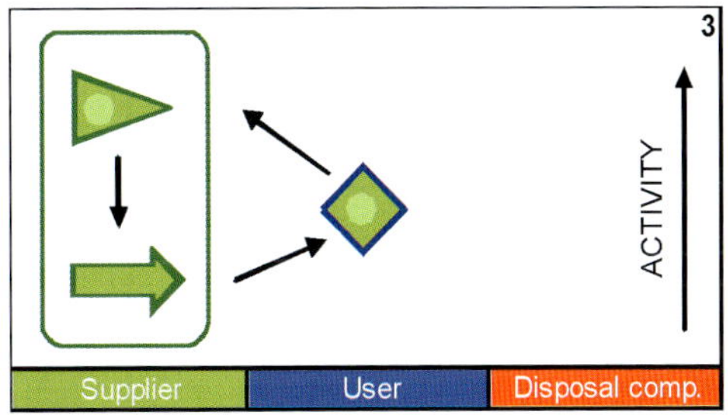

Fig. 5. Business case 3

Cleaning equipment with hydrocarbon solvent
Dr Badawi Chemical Work
GM Egypt
Egypt

GM Egypt is a subsidiary company of the international General Motors group. GM Egypt operates several painting processes which generate VOC emissions and solvent waste. The solvent is used in three main operations: cleaning of guns and connecting piping, sealing and cleaning of tanks. Dr Badawi is supplying the hydrocarbon solvent to GM Egypt, supervises the application of the hydrocarbon containing solvent and regains the solvent waste for recycling. The basis of payment is the amount of produced products per month, which represents a function based charging unit (Fig. 5).

Chemical Leasing in hot dip galvanisation – fluxing process
Zinc Misr
El Sewedy
Egypt

El Sewedy is the Egyptian market leader in hot dip galvanisation industry (Zinc galvanisation), specialised in galvanising towers for power transmission, lighting, communication, etc. Under the Chemical Leasing model, Zinc Misr provides the service for the fluxing process to El Sewedy. Zinc Misr delivers the flux (zinc chloride and ammonium chloride), supervises the process and collects the waste for recycling. The recycling process is not a closed circle because some of the recycled zinc is sold as a different product to someone else. The payment is based on the amount of galvanised products, so it is related to the function of the chemicals (Fig. 6).

	Chemical	Application equipment	Recycling Disposal
Owner	S,U	U	S
Location		U	S
Maintenance		U	S
Operation		U	S
Charging	Service		

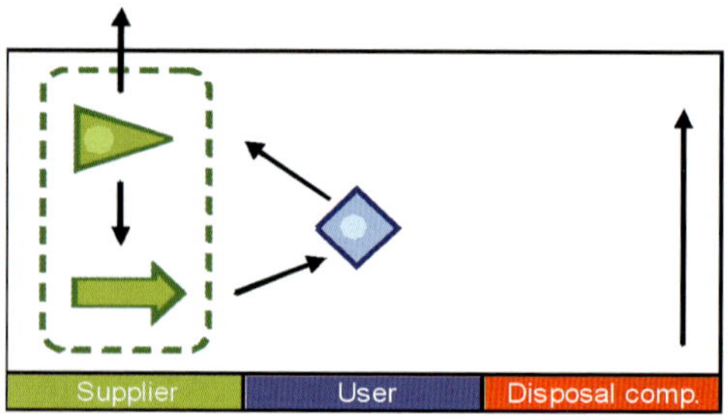

Fig. 6. Business case 4

Sugar mills
Chemical Mac Oil, S.A. de C.V. (Schutz Oil)
Suministro de Materiales Industriales, S.A. de C.V. (SUMAT)
Fideicomiso Ingenio San Cristóbal 80333
Mexico

In this business case there are two suppliers who provide the lubricants to the Mexican sugar mills. The lubricant is delivered to the sugar mill and collected and recycled by the service providers. The machines in the sugar mill are operated by the user. The basis of payment is the amount of sugar cane milled; therefore the unit of charge is based on the function of the used lubricants (Fig. 7).

	Chemical	Application equipment	Recycling Disposal
Owner	S	U	S
Location		U	S
Maintenance		U	S
Operation		U	S
Charging	Service		

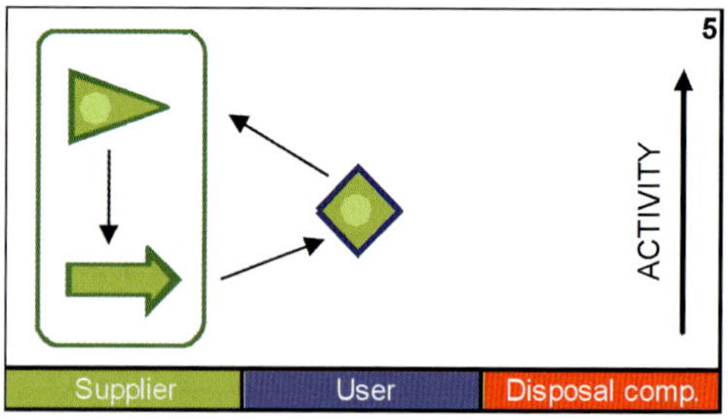

Fig. 7. Business case 5

	Chemical	Application equipment	Recycling Disposal
Owner	S,U	S,U	S
Location		U	S
Maintenance		U	S
Operation		S,U	S
Charging		Service	

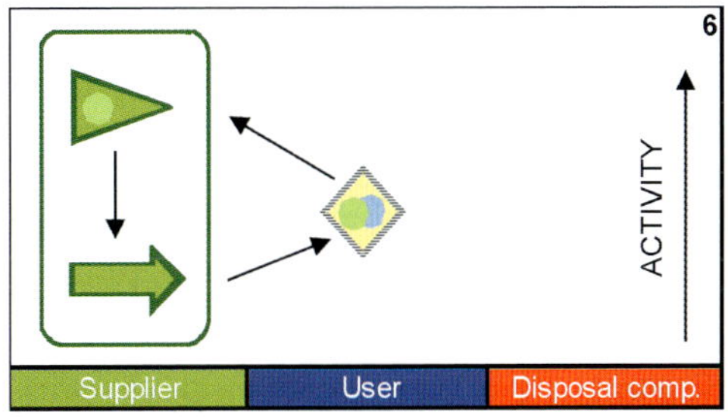

Fig. 8. Business case 6

Electroplating
MARDI Inc., S.A. de C.V.
Cromadora Delgado, S.A. de C.V. (CRODEL)
Mexico

Mardi provides the chemical agent for Crodel's electroplating process. The application equipment in this case is owned and operated by both companies (current supply, dosage for the brighteners), located at Crodel. The recycling is done by Mardi. The unit of payment is the amount of ampere-hours, which corresponds to the surface of the galvanised pieces (Fig. 8).

Glue production wastewater purification
ERG
Henkel-ERA
Russia

ERG has built the water purification plant for Henkel-ERA, which became necessary from the expansion of their glue production. Under the Chemical Leasing regime ERG delivers the chemicals needed by Henkel-ERA. The water is cleaned in a facility owned by Henkel-ERA and operated by ERG, the waste from this process is transported to ERG and treated there. The basis of payment is the amount of purified water, it is thus based on the provided service of water cleaning (Fig. 9).

	Chemical	Application equipment	Recycling Disposal
Owner	S	U	S
Location		U	S
Maintenance		U	S
Operation		S	S
Charging		Service	

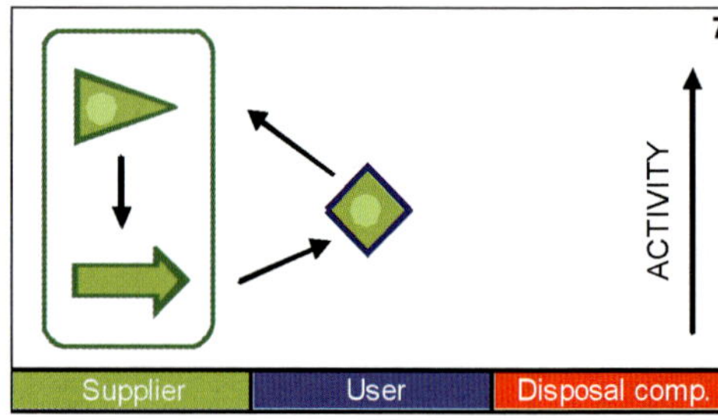

Fig. 9. Business case 7

Metal cleaning
SAFECHEM Europe GmbH
PERO Innovative Services GmbH
MAGNA STEYR Fuel Systems Ges.m.b.H.
Austria

Pero Innovative Services has built a service plant for metal cleaning. The chemicals are provided by Safechem Europe GmbH. Both companies together offer the service to Magna Steyr Fuel Systems Ges.m.b.H.. The total process is managed by the service provider at its location. The basis of payment is the amount of cleaned workpieces (Fig. 10).

	Chemical	Application equipment	Recycling Disposal
Owner	S	S	S
Location		S	S
Maintenance		S	S
Operation		S	S
Charging		Service	

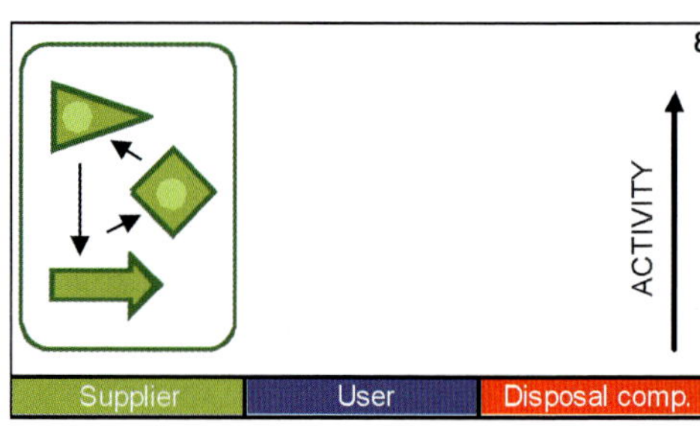

Fig. 10. Business case 8

Summary of the described business cases

In the following table all described cases are summarised. It is clearly shown that in most cases the user is responsible for the application plant and only in the case of Safechem/Pero the supplier cooperation owns, manages and accommodates the plant. The substance is often owned by the supplier, only in 2 cases the ownership changes from supplier to user. The recycling or disposal equipment is managed and owned by the supplier in each of the analysed business cases. Finally, there is clear evidence of economic and environmental benefits gained in all cases where users and suppliers are satisfied with the proposed Chemical Leasing model (Fig. 11).

Business case		**Sub-stance**	**Application plant**				**Disposal plant**			**Unit of charge**
Supplier	User	Owner	Owner	Location	Maintenance	Operation	Owner	Location	Operation	Service or amount
Tiefenbacher	Mepla – Alfit	S	U	U	U	U	S	S	S	Quantity
Akzo Nobel	ABB Arab	S	U	U	U	S	S	S	S	Service
Dr Badawi	GM Egypt	S	U	U	U	U	S	S	S	Service
Zinc Misr	El Sewedy	S,U	U	U	U	U	S	S	S	Service
Schutz Oil et al.	F.I.S.C.	S	U	U	U	U	S	S	S	Service
MARDI	CRODEL	S,U	S,U	U	U	S,U	S	S	S	Service
ERG	Henkel-ERA	S	U	U	U	S	S	S	S	Service
SAFECHEM / PERO	MACNA STEYR	S	S	S	S	S	S	S	S	Service

Fig. 11. Comparison of all business cases

14.3 Evaluation of the Questionnaires

Walter Beyer

All enterprises which participated in the pilot projects were asked for feedback by means of a questionnaire, suppliers and users being questioned separately. In total 9 companies (5 suppliers and 4 users) returned replies, resulting in a response rate of around 30%. The evaluation was carried out separately for suppliers and users. Therefore no statistical evaluation in the narrower sense can be made, but nevertheless trends can be derived.

The results may be summarised as follows:

Was the implementation of the new business model challenging in general?

The definition of a new financial basis for cost charging was considered to be a substantial problem field during the introduction of Chemical Leasing. In all projects the common specifications could be met. However it has to be mentioned that in some cases additional observation periods were assigned in order to check the suitability of the charging basis.
Repliers rated the development of a confidence basis and the formulation of a Chemical Leasing contract as being too tedious.

Would you implement the new business model once again?

It is significant that all companies answered this question with "yes". The results indicate that the business model is interesting for all involved companies despite the difficult phase of the contract establishment and the definition of a new charging basis.

Does the new business model provide cost reductions?
Is the new business model competitive?
Could you use the experiences also for other products/offers?

The presumption that the new business model is competitive and contributes to a cost reduction was also confirmed. Furthermore it was predominantly indicated that this model should be used also for other products.

Consequences of the model

The data concerning the effects of the model reflect the variety of several applications, industries and also the different sizes of the participating enterprises. Nevertheless it shows that the model led to product innovation. The higher quality of customer relationship, which is promoted by the model, is also remarkable. Likewise both groups classify the significance regarding the competition as particularly high.

Is the model also suitable for very small enterprises (5 to 20 employees) or small product quantities?

With the exception of one supplier all indicated that this model is also suitable for small and very small enterprises. Considering the answers to the first question it still has to be examined whether a comprehensive contract design (with the definition of the obligations of the involved parties) and the definition of a new charging basis will be feasible also in very small enterprises, since only rather small quantities of the products are moved.

Which charging base (unit of payment) do you use?

Completely different systems were used to find a suitable charging base. Details can be found in the success stories of Chapters 4 and 5. In summary the information gathered shows a common interest in the business model and its possible meaning for the future, furthermore it shows a common readiness to accept new ways of thinking. In addition, the variety of answers permits the conclusion that the use of auxiliary and operating materials so far had only subordinate significance. Therefore only few reference values for the use of these materials are available.
For the future it is necessary to gather more detailed data about the Chemical Leasing model.

Will this model foster the implementation of REACH?

All except one supplier indicated that this model is supporting the introduction of REACH.
In conclusion it may be summarised that the Chemical Leasing business model represents an interesting additional offer, which results in a distinct decrease of chemicals usage and wastes (above all dangerous wastes). Furthermore, total costs of the user are decreasing on the one hand and on the other hand total profits of the supplier are rising.

List of Contributors

Walter Beyer, Fa. BUK BEYER Umwelt+Kommunikation Unternehmensberatung
Gumpendorferstraße 92/14, 1060 Vienna, Austria
E-mail: office@beyer.at
www.beyer.at

Born 1956 in Vienna, technical education and study of business administration at the Vienna University of Economics and Business Administration; after employment in the banking department of a large company in the steel industry, collaboration with the Austrian Federal Ministry of Agriculture, Forestry, Environment and Water Management in Vienna; working in the area of marketing and PR at an environmental technology production plant; since 1990 independent management consultant. Main areas of specialisation: waste management, integrated management systems for quality – safety – environment, strategy development and development of new business models and ideas, financing and funding; trainer of drivers of dangerous goods, authorised waste management agents and internal auditor; lead auditor for the certification of environmental management systems according to ISO 14001 and EMAS.

Horst Erbel, Managing Director, PERO public limited company
Hunnenstraße 18, 86343 Königsbrunn, Germany
E-mail: pero.erbel.a@pero.ag
www.pero.ag

Horst Erbel was born in Saarlautern in 1936. He concluded two vocational trainings as a technical drawer and as a mechanical engineer. In 1957 he started with the development of his own production of metal cleaning plants. Until then the production of plants had been commissioned externally. In this context he first played the role of the master, then of a production manager in the company of his father P. Erbel. In 1963 he became copartner of the PERO limited partnership, in 1964 he took over the management. In 1979 he became general partner, fully liable for the obligations of PERO limited partnership. In 1992 the Gothaer machinery factory was bought with the supplementary product of continuous-flow washing appliances. In 1996 he transformed the PERO limited partnership into a family-owned public limited company with its headquarters in Königsbrunn and since then he has been managing director of the PERO joint stock company with branch establishments in France (participatory share 49%), Switzerland, and the U.S.A. In 1997 the Gothaer machinery factory was merged with PERO plc. The parent company has 150 staff members and exports 69% of its production. In 2005 he was founder and copartner of the PERO Innovative Services llc, Weiz (Austria).

Ali Hosni, Egypt National Cleaner Production Center (ENCPC)
26 A Sherif Street, Emmobilia Building, 5th Floor, Cairo, Egypt
E-mail: aboohaitham@yahoo.com

Mr. Ali Hosni has academic background in chemistry and geology and about 36 years working experience with the industry. Furthermore, Mr. Hosni has long experience in chemical analysis and quality control laboratories of a Iron and Steel industry, Paint

Industry and Metal industry. Additionally, Mr Hosni has experience in road construction labs and is currently working as a senior CP expert in the Egypt National Cleaner Production Center (ENCPC).

Thomas Jakl, Director Chemicals Policy Unit, V/2, Federal Ministry for Agriculture, Forestry, Environment and Water Management (BMLFUW)
Stubenbastei 5, 1010 Vienna, Austria
E-mail: thomas.jakl@lebensministerium.at

Biologist and earth scientist. Researcher in plant physiology at the University of Vienna until 1991 (specialised in photosynthesis research, simulation of plant bio-productivity). Over 60 publications on environmental issues. 1992: invited speaker at the Alpbach Forum; 1996 winner of the grand Austrian Environment Award (of Lower Austria); top-notch journalist on environmental issues; admission to the "Austrian Scientific Forum for Environment Protection", board member since 2001. Director of the Chemicals Policy Division of the BMLFUW since the end of September 1997; fields of activity: legislation on chemical substances at the national, European and UN levels. Scientific and technical risk assessment of substances and products, best practice diffusion for the production and application of chemical products.

Reinhard Joas, BiPRO GmbH
Grauertstraße 12, 81545 Munich, Germany
E-mail: reinhard.joas@bipro.de

Reinhard Joas studied process engineering and business management, graduating with a doctorate on the economic and ecological aspects of chlorine chemistry. He is the founder and managing director of BiPRO and advises the European Commission, the European Parliament, the Ministries of Environment and Economy as well as business companies and associations on technical, ecological and economic issues.
Reinhard Joas has supervised the activities concerning Chemical Leasing for the Austrian Federal Ministry for Agriculture, Forestry, Environment and Water Management (BMLFUW) as well as for UNIDO from the beginning and is involved in many international projects.

Frank Moser, Associate Programme Officer, United Nations Environment Programme, Chemicals branch
DTIE 11–13, Chemin des Anémones, 1219 Châtelaine – Geneva, Switzerland
E-mail: frmoser@gmail.com

Chemist and Economist. Studies of Chemistry at the University of Konstanz, Germany from 1993 to 1999; main fields of study: biochemistry, organic chemistry; thesis in organic chemistry (polyol peptidomimetics). Studies of Economics at the University of Hagen, Germany from 2001 to 2006; main fields of study: environmental economics, organisation and planning; thesis in environmental economics (Chemical Leasing Business models and the internalisation of negative external effects). Publications on the economic analysis of Chemicals Leasing Business models and their application in developing and developed countries in e.g.: Risk Analysis – An International Journal, Environmental Science and Pollution Research (ESPR). In 2004 and 2005, consultant for the Gesellschaft für Technische Zusammenarbeit (GTZ) and fellow at the United Nations Institute for Training and Research (UNITAR). Since April 2005, Associate Programme Officer at the United Nations Environment Organization (UNEP) Chemical Branch, DTIE in Geneva, Switzerland. Fields of activity: environmentally sound management of Persistent Organic Pollutants (POPs) under the Stockholm Convention on POPs, preparation of guidelines, management of projects in the POPs field.

Ulrich Nagel, TÜV SÜD Management Service GmbH, Head of Certification Body
Ridlerstraße 65, 80339 Munich, Germany
E-mail: ulrich.nagel@tuev-sued.de
www.tuev-sued.de

Ulrich Nagel studied Physics at the Universities of Hanover and Munich from 1968 to 1974. After gaining his Ph.D. at the University of Munich, he joined TÜV SÜD in 1981 as an expert in radiation protection and radioecology. In 1991, he became head of the environmental protection department, focusing on environmental management systems. He qualified as a German government-accredited Environmental Verifier in 1995 and as an ISO 14001 environmental management lead auditor two years later. Head of the Certification Body for environmental management systems from 1996, in 2004 Ulrich Nagel became head of the accredited Certification Body for management systems within TÜV SÜD. In this position, he developed requirements for new management system certifications including Chemical Leasing.
Ulrich Nagel has been a Member of the German Governmental Board of Environmental Management Verifiers since 1999, participating in over one hundred environmental management system certification audits in Germany, France, the United Kingdom, South Africa and the USA.

Cornelia Ohl, UFZ, Helmholtz Centre for Environmental Research – UFZ,
Department of Economics
Permoser Straße 15, 04318 Leipzig, Germany
E-mail: cornelia.ohl@ufz.de

Senior Researcher at Helmholtz Centre for Environmental Research – UFZ, Department of Economics, Leipzig, Germany; Ph.D. 2002, University of Hagen, Germany: Staatliche Umweltregime und transnationales Risikomanagement, Frankfurt/M.: Campus; Board member of the Society for the Advancement of Behavioral Economics (SABE); selected methodological skills: game theory, ecological-economic modelling; selected fields of application: environmental negotiations and global institutions, environmental policy and instruments, publications in e.g.: European Journal of Law and Economics, International Journal of Sustainable Development, Public Choice, Journal of Applied Social Science Studies; selected projects: "Zur Integration risikoökonomischer und strategischer Aspekte bei der ökonomischen Analyse globaler Umweltveränderungen" granted by German Research Foundation (DFG); „Nachhaltige Landnutzung im Spannungsfeld umweltpolitisch konfligierender Zielsetzungen" funded by Federal Ministry of Education and Research (BMBF).

Jorge Pérez, Director, Centro Mexicano para la Producción Más Limpia,
Av. Acueducto S/N, Barrio la Laguna, Colonia Ticoman
Gustavo A. Madero, 07340, Mexico City, Mexico
E-mail: joperez@ipn.mx
www.cmpl.com.mx

Metallurgical Engineering Ph.D. for the Institute of Mining and Technology at the University of New Mexico. Perez has issued more than 60 publications, and he has been professor of Metallurgical Engineering since 1977 and of Postgraduate Programmes since 1985. He has experience in coordination and leadership of equipment. This allowed him to hold the position of President of the Academy of Extractive Metallurgy, Coordinator of the Metallurgical Engineering Postgraduate, Coordinator of the Japanese International Cooperation Agency (JICA) Programme and Director of the Programme of Expansion of Cleaner Production Centres with the Inter-American Development Bank. He has worked as adviser and coordinator in environmental projects of engineering with the national industries and international agencies, among them: JICA, Peñoles Group, PEMEX, Clariant Mexico, Inter-American Development Bank, UNIDO, and others. At this time he is the Director of the Mexican Cleaner Production Centre (CMP+L).

Christian Plas, denkstatt GmbH,
Hietzinger Hauptstraße 28, 1130 Vienna, Austria
E-mail: christian.plas@denkstatt.at
www.denkstatt.at

Christian Plas, Ph.D., Managing Director of denkstatt, graduated from the University of Natural Resources and Applied Life Sciences, Vienna, with a degree in Environmental Engineering and Water Management and a Ph.D. in Biotechnology. In 1993 he founded denkstatt with offices in Vienna, Budapest and Timisoara focusing on sustainable development and environmental management in the industrial sector. Since 1992 he has been member of the Environmental Policy Committee of the Federation of Austrian Industry and of diverse temporary committees of the Federation of Austrian Industry and the Austrian Federal Economic Chamber. Regular seminars in environmental planning and management, CO2-management and risk management.

Ignacio García Sánchez, Centro Mexicano para la Producción Más Limpia,
Av. Acueducto S/N, Barrio la Laguna, Colonia Ticoman
Gustavo A. Madero, 07340, Mexico City, Mexico
E-mail: igarcias@ipn.mx

Graduated in Oceanography Engineering at the "Universidad Autonoma de Baja California (UABC)", specialised in Resources Management, M.Sc. Environmental Eng. by the ITESM, Monterrey. Studies on Occupational Safety and Health Assessment, Environmental Risk Assessment, Engineering Design, Financial Analysis and Business Planning. Publications in research journals, such as "Environmental Quality", "AWMA", many technical and humanistic courses. Has been involved in the environmental area, in the design of environmental plans, in environmental audit, power efficiency and industrial security. Since 2003 he has been working as Technological Development Engineer at the research department of the Mexican Cleaner Production Centre, providing technical support for industrial programmes (environmental prevention and energy, in many sectors like: sugar, oil, chemical plants, electroplating, food processing, fisheries, among other areas. Since June 2006 he has been a Chemical Leasing Project Manager in Mexico.

Peter Schaff, TÜV SÜD PEOPLE Business Unit
Ridlerstraße 65, 80339 Munich, Germany
E-mail: peter.schaff@tuev-sued.de

Prof. Dr. Dr. med. habil. Peter Schaff studied medicine in Munich and mechanical engineering/biomechanics in the USA, completing his professorial thesis on the subject of applied biomechanics. Already during his studies, Schaff held various positions, for example at the Max-Planck-Institute in Munich. After practising as medical doctor, one year in the USA and working for PEAK Performance Inc., he joined today's TÜV SÜD in 1988, where he started his career as head of the Biomechanics Institute of TÜV Product Service GmbH. In 1995, Prof. Schaff was appointed manager of the medical, health and sports unit at TÜV SÜD Product Service, where he assumed further responsibility in 1998 as Prokurist (holder of a general commercial power of attorney) and member of the management team. Peter Schaff has been Managing Director of TÜV SÜD Management Service GmbH since 2000 and head of the PEOPLE Strategic Business Segment at TÜV SÜD since 2005. In addition to TÜV SÜD Management Service, the PEOPLE Strategic Business Segment also includes TÜV SÜD Life Service and TÜV SÜD Akademie. During his activities at TÜV SÜD, Prof. Schaff has also been involved in a host of research projects and has regularly acted as lecturing professor at the Munich Technical University since 2003.

Rudolf Schott, AFC Aforma Consult, Unternehmensberatung GmbH
Rauscherstraße 15/14, 1200 Vienna, Austria
E-mail: rudolf.schott@afc-consult.at

Education in engineering at the College for Technical Chemistry (HBLVA), Vienna; sworn and court-certified expert for waste and for the production of paints and varnishes. After 15 years of industrial experience, he founded AFC Consult in Vienna in 1993 and acts as consultant for Austrian companies, national and international industrial associations, Austrian ministries, and for the European Commission.

Petra Schwager, UNIDO Cleaner Production Programme, PTC/ECB Branch,
Vienna International Centre
P.O. BOX 300, 1400 Vienna, Austria
E-mail: p.schwager@unido.org
www.unido.org/cp

Petra Schwager works for the Energy and Cleaner Production Branch of the United Nations Industrial Development Organization (UNIDO). In 2004 she initiated UNIDO's Chemical Leasing activities. Since 1995 she has been involved in building-up UNIDO's global Cleaner Production Programme, which currently includes 38 countries. She has managed National Cleaner Production Centres and Programmes in more than 15 countries and developed UNIDO's Regional Cleaner Production Network for Latin America.

Fostering Public-Private Partnerships and encouraging innovative approaches to sustainable chemicals management in developing and transition countries are additional fields of her work. She studied Economics at the Vienna University of Business Administration and Social Sciences and Environmental Management at the University of London.

Ali Abo Sena, National Expert on Chemical Leasing – Project Manager,
Ministry of Trade and Industry,
UNIDO – ENCPC Egypt National Cleaner Production Centre
26A Sherif Street, Immbolia Building, 5th Floor, Downtown, Cairo, Egypt
E-Mail: ali.abosena@mfti.gov.eg, abosena@yahoo.com

Ali Abo Sena has long standing experience in the field of environmental protection. In 1996 he graduated from the Faculty of Science at Mansoura University, Egypt, with B.Sc. in Environmental Sciences (Major Environmental Chemistry). Abo Sena has a Diploma in Environmental Management for Developing Countries (2002, Dresden University of Technology, Germany) and Master of Science in Environmental Management and Policy (2004, Lund University, Sweden). He was awarded the SYSAV prize for the best thesis on waste management in Sweden in 2004. In addition to about 2 years of industrial experience (dairy company, consulting company), Abo Sena has about 5 years work experience as an Environmental and Industrial Inspector (Group Leader) at the Egyptian Environmental Affairs Agency (EEAA). Currently, Abo Sena is working for the United Nations Industrial Development Organization (UNIDO) – Egypt National Cleaner Production Centre as a coordinator for the Chemical Leasing project in Egypt.

Alexander Startsev, North Western International Cleaner Production Centre
Zagorodny prospect 8, off. 15, 191002 St. Petersburg, Russian Federation
E-mail: a.startsev@nwicpc.ru
www.nwicpc.ru

Alexander Startsev has a degree in risk management of the St. Petersburg University of Telecommunications and a law degree (environmental law) of the St. Petersburg Marine Transport University. In addition Startsev holds a diploma of the Business School at the International Management Institute of St. Petersburg (IMISP). From 1990 – 2000 he was Director General of the St. Petersburg Ship-Breaking Yard (Scrap Metal Processing Plant) "Interferrum-Metal Ltd." Currently Alexander Startsev is Director General of the UNIDO North-Western International Cleaner Production and Environmental Management Centre (NWICPC, UNIDO project for North-West Russia), St. Petersburg, which deals among others with CP-investment promotion and EST promotion. Under the management of Alexander Startsev, NWICPC has initiated preparations for the Chemical Leasing project in 2004. Recently a group of 5 national experts in Chemical Leasing has been trained and the Chemical Leasing project work is been currently carried out with 7 companies as users in North-West Russia. At the same time Startsev is an Assistant to the Chairman of the Environmental Committee of the Russian Federation State Duma (Russian Parliament), Moscow, on a honorary basis. He is also the National Expert of the Higher Ecological Council attached to the Environmental Committee of the Russian Federation State Duma.

Richard Tuschl, Institute for Industrial Ecology
Tor zum Landhaus – Rennbahnstraße 29/C/3, 3100 St. Pölten, Austria
E-mail: richard.tuschl@noe-lak.at

Studies in Informatics as well as in IT-Management at the Technical University of Vienna and the University of Vienna. Graduated from the Technical University of Vienna with a thesis on matrix methods in the evaluation of products under the aspect of sustainability. Establishment of an enterprise in the field of Management Consultancy and IT in 2000. Since 2005 scientific staff member at the Institute for Industrial Ecology.

Elisa Arreola Valerio, Centro Mexicano para la Producción Más Limpia,
Av. Acueducto S/N, Barrio la Laguna, Colonia Ticoman
Gustavo A. Madero, 07340, Mexico City, Mexico
E-mail: earreolav@ipn.mx
www.cmpl.com.mx

Escuela Superior de Ingeniería Química e Industrias Extractivas – IPN. 2001 Ingeniero Químico Industrial. Experiencia: 2005 – 2006 Mexican Cleaner Production Centre, Project Engineer, responsible for the quality management system ISO 9001:2000 and coordinator in charge for the development of environmental management ISO 14001:2004. Development of the life cycle assessment project. Development of the Chemical Leasing project. Recognition visits to companies. Implementation of the CMP+L with the industry. Courses coordination. Development and implementation of the consultants certification.

Andreas Windsperger, Institute for Industrial Ecology
Tor zum Landhaus – Rennbahnstraße 29/C/3, 3100 St. Pölten, Austria
E-mail: andreas.windsperger@noe-lak.at
www.indoek.noe-lak.at

Scientific director of the Institute for Industrial Ecology. After diploma-studies at the University of Technology and at the University of Agriculture in Vienna he graduated with a diploma in Biotechnology and Bioengineering in 1980. After further studies he received his Ph.D. degree from the University of Technology, Vienna, with a thesis on Bioengineering in Environmental Technology in 1985. Since 1990 he has been Associated Professor for Chemical Reaction Engineering at the University of Technology, Vienna. From 1990 up to 1993 he spent almost four years in industry and was responsible for environmental planning and engineering in a chemical factory. During that time he was responsible for the operation of the environmental plants as well as for environmental strategies and reporting in the company. For that purpose he established the appropriate organisation in his team (about 20 persons) which guaranteed efficient service for the entire company. As of 1992 the department worked also for external clients with a license for environmental engineering. From 1994 to 1998 he headed the Research Institute for Chemistry and Environment at the Vienna University of Technology, which has been continued as the Institute for Industrial Ecology at the Academy of Lower Austria since July 1998.

Michael Wittmann, Federal Ministry for Agriculture, Forestry, Environment and Water Management (BMLFUW), Legal Expert Chemicals Policy Unit, V/2
Stubenbastei 5, 1010 Vienna, Austria
E-mail: michael.wittmann@lebensministerium.at
www.lebensministerium.at

Michael Wittmann studied law at the University of Vienna and held several positions in Austrian ministerial offices, specialising in environmental law. As legal expert of the Chemicals Policy unit, he deals with national, European and international chemicals law. In 2002, he co-edited the publication "The Role of Precaution in Chemicals Policy".
A self-taught artist, he draws caricatures on the spot and illustrations for a major German Science Fiction serial.

Institute for Entrepreneurship and Innovation
Vienna University of Economics and Business Administration
Nordbergstraße 15, 1090 Vienna, Austria
E-mail: entrepr-sekr@wu-wien.ac.at
www.wu-wien.ac.at/inst/entrep/

The students named below are majoring in "entrepreneurship and innovation management" under the direction of Univ.-Prof. Dr. Nikolaus FRANKE at the Vienna University of Economics and Business Administration. The compilation of the business plan "Chem-Kit" in particular was assisted by the course instructors Mag. Rudolf *Dömötör* and Dr. Martin *Schreier.*

Sissi Chen

Sissi is a student of "international business administration and sinology". Due to her passion for travelling and international business she has already completed several internships at various companies located in China to gain work experience. She has been a professional figure skater for more than eight years and has also taken part in international competitions.
E-mail: h0255216@wu-wien.ac.at

Zin Duong

Along his studies at an Austrian school for tourism management Zin was able to gain a lot of international work experience at authorised tourist operations and certified travel-related business companies, amongst others in Germany, Great Britain and Qatar. Currently he is majoring in both "entrepreneurship and innovation management" and "international business". His main interests focus on future business models and opportunities complying with social responsibility on an international basis, in particular for the Chinese market where his own roots are situated.
E-mail: h0252837@wu-wien.ac.at

Margarete Mayerhofer

Margarete studied pharmacy at the Faculty of Natural Sciences, University of Vienna, from 1995 to 2001. In her master thesis she engaged in the synthesis of cytostatic drugs. After graduation, she underwent professional training to be pharmacist. In addition to working at a pharmacy, she studies at the Vienna University of Economics and Business Administration.

E-mail: h9502130@wu-wien.ac.at

Reinhild Messner

Reinhild, born in Vienna in 1984, studies "business-economics education and socioeconomics" at the Vienna University of Economics and Business Administration. As a former member of the European Youth Parliament and a student of the programme for "entrepreneurship and innovation management", Reinhild is highly motivated to contribute to a safe and sustainable future. Co-authoring a proposal for a model consulting agency specialised in Chemical Leasing gave her the unique possibility to demonstrate her considerable social and environmental commitment.

E-mail: h0352009@wu-wien.ac.at

Andreas Josef Muhr

Andreas is a student of Business Administration at the Vienna University of Economics and Business Administration. He is majoring in "entrepreneurship and innovation management" as well as "management information systems". His minors include Project Management. Andreas' varied professional experience comprises working at inno group, Accenture and Eastman Kodak as well as co-founding his own consulting firm at the age of twenty. The Upper Austrian pursues a strong interest in finding and exploiting business opportunities, especially in the dynamic, ever-growing field of efficient raw materials management.

E-mail: h9851247@wu-wien.ac.at

Eva Maria Riebenbauer

Eva is studying International Business Administration with majors in "entrepreneurship & innovation management" and "international human resource management". Another focus within her field of studies represents "globalisation and regional development". Over the last years she gained professional experience through her work in the sales department of Siemens Building Technologies. In summer 2006 she spent an exchange semester at the Dauphine University in Paris.

E-mail: h0251212@wu-wien.ac.at

Springer and the Environment

We at Springer firmly believe that an international science publisher has a special obligation to the environment, and our corporate policies consistently reflect this conviction.

We also expect our business partners – printers, paper mills, packaging manufacturers, etc. – to commit themselves to using environmentally friendly materials and production processes.

The paper in this book is made from no-chlorine pulp and is acid free, in conformance with international standards for paper permanency.